Lie Groups

A Problem-Oriented Introduction via Matrix Groups

About the cover

The cover is a metaphoric representation of the Lie group

$$G = \left\{ \begin{bmatrix} 1 & a & b \\ 0 & 1 & a \\ 0 & 0 & 1 \end{bmatrix} : a, b \in \mathbb{R} \right\}$$

and its Lie algebra

$$L(G) = \left\{ \begin{bmatrix} 0 & u & v \\ 0 & 0 & u \\ 0 & 0 & 0 \end{bmatrix} : u, v \in \mathbb{R} \right\}.$$

It also shows the two curves (one parameter subgroups)

$$\gamma_1(t) = \begin{bmatrix} 1 & t & t^2/2 \\ 0 & 1 & t \\ 0 & 0 & 1 \end{bmatrix} \quad \text{and} \quad \gamma_2(t) = \begin{bmatrix} 1 & 0 & t \\ 0 & 1 & 0 \\ 0 & 0 & 1 \end{bmatrix}$$

passing through the identity I of G, along with their associated tangent vectors at the identity

$$A_1 = \gamma_1'(0) = \begin{bmatrix} 1 & 1 & 0 \\ 0 & 0 & 1 \\ 0 & 0 & 0 \end{bmatrix} \quad \text{and} \quad A_2 = \gamma_2'(0) = \begin{bmatrix} 0 & 0 & 1 \\ 0 & 0 & 0 \\ 0 & 0 & 0 \end{bmatrix}$$

in $L(G)$. Note that $\gamma_j(t) = \exp(tA_j)$ and $tA_j = \log(\gamma_j(t))$ for $j = 1, 2$ and t in $\mathbb{R}$. The group G is a subgroup of the Heisenberg group $\mathcal{H}$ (see problem 5.1.9 in section 5.1).

Cover design by Elizabeth Holmes Clark of blue oak design (www.bluoak.com)

Library of Congress Catalog Card Number 2009933070

Print edition ISBN: 978-0-88385-759-5

Electronic edition ISBN: 978-1-61444-604-0

Printed in the United States of America

Current Printing (last digit):
10 9 8 7 6 5 4 3

Lie Groups

A Problem-Oriented Introduction via Matrix Groups

Harriet Pollatsek
Mount Holyoke College

Published and distributed by
The Mathematical Association of America

MAA TEXTBOOKS

Calculus Deconstructed: A Second Course in First-Year Calculus, Zbigniew H. Nitecki
Combinatorics: A Problem Oriented Approach, Daniel A. Marcus
Complex Numbers and Geometry, Liang-shin Hahn
A Course in Mathematical Modeling, Douglas Mooney and Randall Swift
Cryptological Mathematics, Robert Edward Lewand
Differential Geometry and its Applications, John Oprea
Elementary Cryptanalysis, Abraham Sinkov
Elementary Mathematical Models, Dan Kalman
Essentials of Mathematics, Margie Hale
Field Theory and its Classical Problems, Charles Hadlock
Fourier Series, Rajendra Bhatia
Game Theory and Strategy, Philip D. Straffin
Geometry Revisited, H. S. M. Coxeter and S. L. Greitzer
Graph Theory: A Problem Oriented Approach, Daniel Marcus
Knot Theory, Charles Livingston
Lie Groups: A Problem-Oriented Introduction via Matrix Groups, Harriet Pollatsek
Mathematical Connections: A Companion for Teachers and Others, Al Cuoco
Mathematical Interest Theory, Second Edition, Leslie Jane Federer Vaaler and James W. Daniel
Mathematical Modeling in the Environment, Charles Hadlock
Mathematics for Business Decisions Part 1: Probability and Simulation (electronic textbook), Richard B. Thompson and Christopher G. Lamoureux
Mathematics for Business Decisions Part 2: Calculus and Optimization (electronic textbook), Richard B. Thompson and Christopher G. Lamoureux
The Mathematics of Games and Gambling, Edward Packel
Math Through the Ages, William Berlinghoff and Fernando Gouvea
Noncommutative Rings, I. N. Herstein
Non-Euclidean Geometry, H. S. M. Coxeter
Number Theory Through Inquiry, David C. Marshall, Edward Odell, and Michael Starbird
A Primer of Real Functions, Ralph P. Boas
A Radical Approach to Real Analysis, 2nd edition, David M. Bressoud
Real Infinite Series, Daniel D. Bonar and Michael Khoury, Jr.
Topology Now!, Robert Messer and Philip Straffin
Understanding our Quantitative World, Janet Andersen and Todd Swanson

MAA Service Center
P.O. Box 91112
Washington, DC 20090-1112
1-800-331-1MAA FAX: 1-301-206-9789

To Sandy, Shura and David

Preface

Two things are distinctive about this book: the low prerequisites (multi-variable calculus and linear algebra) and the development via problems (nearly 200 are provided). This is a text for active learning and one that can be used by students with a range of backgrounds. The material can be covered in a single semester.

In 1976–77, at the University of Oregon on my first sabbatical, I sat in on a course on Lie groups taught by Richard Koch. Now, I had studied Lie algebras and groups of Lie type as a graduate student (I am an algebraist steeped in the finite classical groups and their geometries), so I knew how to start with Lie algebras and define groups, but Koch's course was a revelation to me: the groups came first! I became so excited that I was determined to teach a course to Mount Holyoke undergraduates that would expose them to some of these beautiful ideas. My department indulged me, and the following year I tried. It was a dismal flop. I spent an entire semester trying to help my students understand left-invariant vector fields on a manifold. I learned a lot, even if perhaps they didn't. I realized that I needed to suppress the differential geometry and topology and start with matrix groups. I wasn't the only one who thought so. In 1979, Morton Curtis published *Matrix Groups*, although he included topology. In 1983, Roger Howe made an eloquent argument that Lie groups belong in the undergraduate curriculum in "Very Basic Lie Theory" in the *American Mathematical Monthly*. But no one took as extreme a view as I did: trying to create a course that students could take right after multi-variable calculus and linear algebra. (This is no longer quite true: John Stillwell's *Naive Lie Theory*, published in 2008, has similar prerequisites but is aimed more specifically at juniors and seniors.)

Why so extreme? At that time, my entire department was having conversations about the role of prerequisites and our desire to bring more advanced topics into the undergraduate curriculum. We had two concerns. One was that our students were choosing a major before they had seen much of the breadth and richness of our discipline. Another was that, since most of our majors did not go on to graduate study in mathematics, most of our students *never* saw many important mathematical ideas. We were audacious enough to propose to the U.S. Department of Education's Fund for the Improvement of Post-Secondary Education (FIPSE) the creation of seven reduced-prerequisite courses on advanced topics, and FIPSE was bold enough to fund us to make them happen. The FIPSE funding not only paid for release time for course developers, it also — perhaps more importantly— funded a series of sabbatical visitors who sat in on and critiqued the first offerings of our courses. In my own case, I was helped by Ken Rogers of the University of Hawaii, who sat in on an early matrix group version. More recently, I benefitted from comments by David Murphy, who used my materials in a Linear Algebra II course at Kalamazoo College.

In order to make this material accessible to students who have taken just multivariable calculus and linear algebra, I have concentrated on the algebraic ideas, with just enough analysis to define the tangent space and the differential and to make sense of the exponential map. Topology is excluded except for informal references to connected and simply connected sets and to closed sets. I also exclude quotient structures, except in asides for students who have studied abstract algebra. On the other hand, I do include a brief discussion of groups of Lie type over finite fields, the algebraist's Lie algebras first and groups second approach, very beautiful in its own right.

Over the years I've taught Lie groups many times at Mount Holyoke, and the enthusiasm and achievements of my students have consistently inspired me. This text, consisting mostly of problems, represents the form my course took when it stabilized. It's always true that students learn best when they are actively engaged, but it seems to me that when mathematically inexperienced students encounter something very different, it's especially important that they work through the new ideas for themselves. When I teach this material I lecture very little; most class time is spent with students either working on the problems or presenting and discussing their solutions. Sometimes, especially when we begin a new topic, my students work on problems all together, with me as writer at the board; other times, they work in pairs or in small groups.

My students are usually quite diverse in preparation, and often a number of science majors are among them. (In fact, the most recent time I taught Lie groups, just half the students in my class were math majors.) Many of my students have had only a very mild exposure to reading and writing formal proofs. The workshop-like format permits me to offer these students the extra support that they need. This text includes many routine problems, which my less experienced students have found helpful, as well as a good number of more challenging ones. In particular, a few problems ask students to "explain why" a statement is true for a statement that, with more sophisticated students, one would simply regard as obvious. Some problem sequences repeat very similar ideas, giving these students extra practice.

What about students who have already studied abstract algebra or real analysis? My experience has been that these students are also both engaged and challenged by this material. Obviously I steer them to the more substantial problems, and I often suggest supplementary reading. When students work in groups during class, I'm mindful of the students' backgrounds in helping them form groups that will function well. On the other hand, students who go on to abstract algebra after Lie groups (and some who hadn't originally intended to go on in fact do) carry with them a rich store of concrete examples to motivate and illuminate the ideas they encounter when systematically studying groups, rings and fields. In particular, my students acquire a deep understanding of the concept of an algebraic structure and its morphisms. Perhaps more important, they also develop an appreciation of the inter-connectedness of mathematics and the sometimes surprising ways in which it provides a language to describe and understand the physical world.

Because so much of the content of the book is scattered among the problems, every chapter has a section called *Putting the pieces together* in which all definitions and results are collected for reference. More advanced theorems of which the problems are special cases and references to other applications are sometimes given in this section as well. It also includes suggestions for further reading.

One of my favorite books is *Introduction to Affine Group Schemes* by William Waterhouse (Graduate Texts in Mathematics, Springer-Verlag, 1979). One of the (many) things I admire about it is the way the author occasionally includes a "vista" that puts the text material in a larger context. I've adopted this strategy with a section at the end of every chapter called *A broader view.* Sometimes the context is mathematical and sometimes historical (or both). These sections are not detailed, but I hope they give a flavor of the terrain the reader glimpses.

Appendix I includes the specific facts from linear algebra that I expect students to know and be able to use. I assume they will consult their own linear algebra books to fill any gaps.

There is now a small but growing body of material that students at this level can profitably read. Some of it I assign for supplementary reading. I also require my students to write a short expository paper based on reading about something not covered in class or in the text. This is another opportunity to tailor the assignment to students' different levels of preparation. The problems in Chapter 6 are less demanding than in the other chapters, so students have time to work on their papers. In fact, I treat the material in Chapter 6, especially in 6.2, largely as an opportunity to review. Appendix II includes my paper assignment and list of possible topics.

Finally, Appendix III includes several suggestions for more extensive reading, building on some of the ideas in this text. Here I hope the interested student (or instructor) will find ideas for a second semester of study, and students with more background will find enriching and challenging collateral reading.

The Bibliography includes the sources in Appendices II and III, in addition to those cited in the text or mentioned as suggested readings.

Notational Conventions

Throughout the text, x.y denotes section y in Chapter x (referred to as just section y within Chapter x). The label (x.y.z) means problem z in x.y. The notation **(S)** following a problem number (or letter) means that a solution appears in the back of the text. (The solutions to all problems are in a supplement for instructors.) A few of the most challenging problems are marked **(C)**. I use the word "result" to mean a theorem, proposition, corollary or lemma. Ordinarily, Result x.y means result y in Chapter x. An exception is in Chapter 6: the results in the (fifth) section *A broader view* are numbered Result 6.5.x for result x in 6.5.

An asterisk on a result number means its proof is not obtainable using the elementary methods in this book. With few exceptions, the proof of an un-starred result is either in the problems in that chapter or is sketched in the text. Proofs or sketches of proofs are concluded with the symbol $\square$. The symbol $\Leftrightarrow$ means "if and only if." I use the symbol $\mathbb{R}$ for the set of all real numbers, $\mathbb{Z}$ for the integers and $\mathbb{Q}$ for the rational numbers. Ordinarily, matrices are denoted by capital letters (A), vectors by boldface letters ($\boldsymbol{a}$), and scalars by lower case (a). Linear transformations are typically denoted by capital letters also (T). Other functions are represented in a variety of ways, depending on context.

Contents

CHAPTER 1

Symmetries of vector spaces

1.1 What is a symmetry?

For mathematicians, *symmetries* are special invertible functions. The language comes from the idea of symmetry of a geometric object, say a polygon made out of a sheet of paper lying on a flat surface. To use the informal language of grade school mathematics, a symmetry of a polygon is a motion that corresponds to picking up the paper polygon and replacing it so that it appears unmoved. We say the motion *preserves* the polygon. (In this informal language, we assume that both sides of the paper look the same, and also that the various parts of the polygon are unmarked.)

To be specific, consider symmetries of a square. Imagine a paper square placed with its center at the origin of the plane $\mathbb{R}^2$ and with its sides parallel to the coordinate axes. (See Figure 1.1.) If the square is rotated counter-clockwise around its center (the origin) through an angle of $\pi/2$, it will again line up with its original position and will appear unmoved. So this rotation is a symmetry of the square. Similarly, a rotation of the square around the origin, clockwise or counter-clockwise, through any integer multiple of $\pi/2$ is a symmetry of the square.

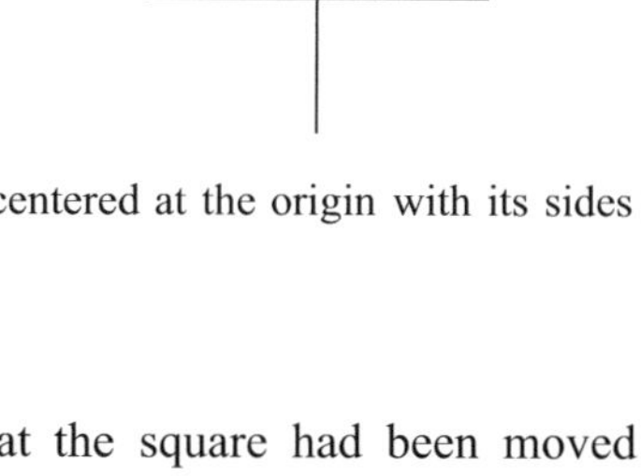

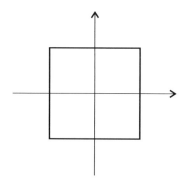

Figure 1.1. A paper square centered at the origin with its sides parallel to the coordinate axes

Problem 1.1.1.

a. Could you recognize that the square had been moved if it were rotated counter-clockwise about the origin through an angle of $\pi/4$? Draw a picture to illustrate.

h. Compare the effects on the square of the rotation counter-clockwise about the origin through an angle of $\pi/2$ and the rotation clockwise about the origin through an angle of $3\pi/2$. Draw pictures that include labels on the vertices of the square to illustrate your thinking.

Two symmetries (or, more generally, two functions) are the same (i.e., they are **equal**) if they have the same effect on their common domain. For example, a rotation of the square through $\pi/2$ counter-clockwise about the origin and a rotation through $3\pi/2$ clockwise about the origin are the same symmetry of the square, even though their descriptions are different. With this in mind, the paper square turns out to have a total of four rotation symmetries. This count includes the rotation through an angle of zero — in other words, it includes the "motion" that leaves the square exactly where it was.

Problem 1.1.2.
Give a geometric description of each of the four different rotation symmetries of the paper square.

Recall that a function $T: X \to Y$ is **invertible** if there is a function $S: Y \to X$ which "undoes" T in the sense that $S(T(P)) = P$ and $T(S(Q)) = Q$ for every P in X and every Q in Y. If T is invertible, we write $S = T^{-1}$, the **inverse** of T.

Problem 1.1.3. (S)
Let T_1 be the rotation of the square counter-clockwise around the origin through an angle of $\pi/2$. Is T_1 invertible? To answer this question, describe a motion S_1 of the square which "undoes" T_1, in the sense that doing T_1 and then S_1 puts the square back where it started, and similarly doing first S_1 and then T_1 also puts the square back where it started.

The x-axis divides the paper square "symmetrically" in half. In fact, your first exposure to the use of the word *symmetric* might have been in this context — the square is symmetric across the x-axis. There are exactly four lines that divide this square symmetrically in half.

Problem 1.1.4.
What are the other three lines that divide this paper square symmetrically in half?

For each of the special lines that divides the paper square symmetrically in half, there is a corresponding symmetry of the square consisting of "flipping" (or *reflecting*) the square over the line. It turns out that these four reflections together with the four rotations described in (1.1.2) constitute *all* of the symmetries of this paper square.

Problem 1.1.5.
Consider a non-square paper rectangle centered at the origin of $\mathbb{R}^2$ and with its sides parallel to the coordinate axes. (See Figure 1.2.)

a. **(S)** Which of the 8 symmetries of the paper square are also symmetries of this rectangle? How many symmetries does this non-square rectangle have?

b. We say one figure is "more symmetrical" than another if it has more symmetries. Which is more symmetrical, the square or the non-square rectangle?

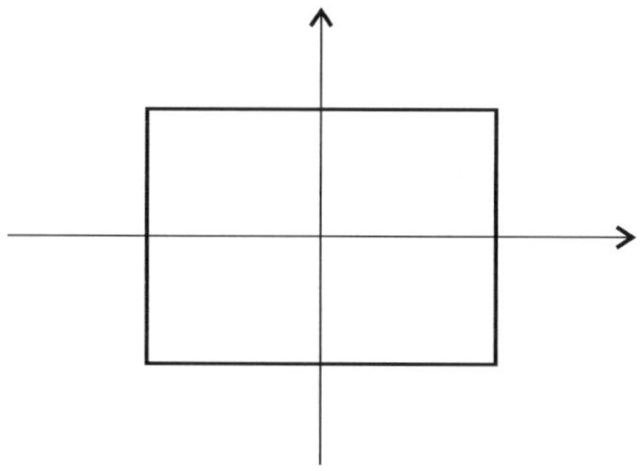

Figure 1.2. A paper non-square rectangle centered at the origin with its sides parallel to the coordinate axes

c. Still referring to paper polygons, how many symmetries does an equilateral triangle have? an isosceles triangle which is not equilateral? a triangle which is not isosceles? Which kind of triangle is most symmetrical?

Problem 1.1.6.

Consider a circular disk of paper centered at the origin of $\mathbb{R}^2$. (See Figure 1.3.)

a. Which rotations around the origin are symmetries of the disk?

b. Which lines divide the disk symmetrically in half?

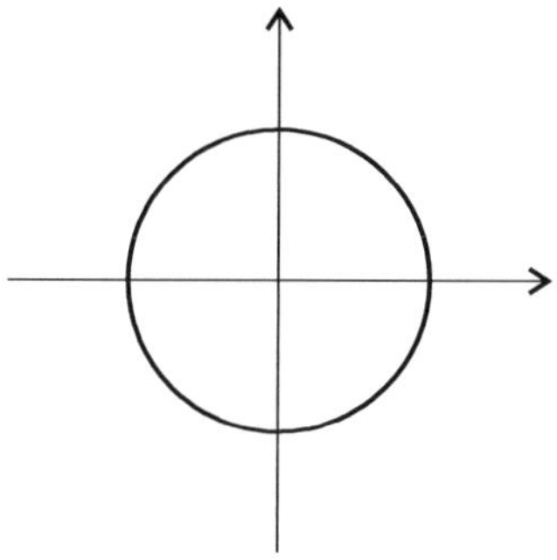

Figure 1.3. A circular disk of paper centered at the origin

Remark Mathematicians say that the circular disk in (1.1.6) has "continuous" symmetries, in the sense in which the real line $\mathbb{R}$ is continuous. In contrast, the paper polygons in the preceding problems have "discrete" symmetries, in the sense in which, for example, the set of numbers $\{0, \pi/2, \pi, 3\pi/2\}$ is discrete—there are gaps between the numbers. Note that an infinite set can be discrete in this sense; for example, the set of integers is discrete. The focus in this book will be on continuous symmetries.

Consider the function $T_2\colon \mathbb{R}^2 \to \mathbb{R}^2$ given by $T_2(x, y) = (x, -y)$. This function is a *reflection* of the plane across the x-axis. Calling T_2 a "reflection" is an adaptation to two dimensions of the idea of reflecting an object in a mirror in three dimensions. In three dimensions, an input point P and its mirror image P' are equidistant from the mirror, measured perpendicular to the mirror. (Both P and its mirror image P' are regarded as "real" points in three-dimensional space, as if the mirror were a clear glass plane dividing space.) In this two-dimensional example, an input point $P = (x, y)$ and its mirror image $P' = (x, -y)$ are equidistant from the x-axis, measured perpendicular to the x-axis. The x-axis is called the *mirror line* of the reflection.

Problem 1.1.7. **(S)**

a. How do you know that the reflection of the paper square across the x-axis is invertible? Answer this by describing what you would do to the paper square to undo this reflection.

b. Now consider the reflection $T_2 \colon \mathbb{R}^2 \to \mathbb{R}^2$ described above. Show that T_2 is invertible by finding an explicit formula for its inverse T_2^{-1}. Using your formula, verify that $T_2^{-1}(T_2(x, y)) = (x, y)$ and $T_2(T_2^{-1}(x, y)) = (x, y)$ for all (x, y) in $\mathbb{R}^2$.

c. In two dimensions, every reflection T fixes each of the points on its mirror line (i.e., $T(P) = P$ for any point P on the mirror line of T). Explain why this is true for the reflection of the paper square across the x-axis, using your geometric description. Then verify it for the reflection $T_2 \colon \mathbb{R}^2 \to \mathbb{R}^2$ using the formula for T_2.

d. Does T_2 fix any points *not* on its mirror line? Is this true in general for reflections? Explain.

Recall that a function $T \colon X \to Y$ is invertible if and only if it is both one-to-one and onto. The terms injective and surjective are also used. A function is **one-to-one** (*injective*) if distinct input values give distinct output values. Usually we verify that a function T is one-to-one by checking the *contrapositive* of this statement: we assume that $T(P) = T(Q)$ for some P and Q in the domain X of T and then show that $P = Q$. A function T is **onto** the set Y (*surjective*) if given any P' in Y, we can find a P in X with $T(P) = P'$.

Problem 1.1.8. (S)

Verify that the function $T_2 \colon \mathbb{R}^2 \to \mathbb{R}^2$ described above is both one-to-one and onto.

The length of the side of the paper square in our first example was definite, but never specified. We could take a different point of view. Suppose we regard *every* square centered at the origin with sides parallel to the axes as "the same." (This isn't a crazy idea. When you see an object from up close it looks larger than when you see it from far away, but your brain can recognize the object as "the same" in both views.) Now, we can change what we mean by a symmetry to encompass this new notion of "sameness." In this new context, we can think of our square as made of a miraculous elastic substance. The symmetries of this square now include more than just the earlier rotations and reflections. In addition, we can stretch or shrink the square, as long as we do so uniformly, so it remains a square centered at the origin and having its sides parallel to the axes. Under this new definition, our square has many more than eight symmetries.

Problem 1.1.9.

Fix $a > 0$. How do you know that the function $T_a \colon \mathbb{R}^2 \to \mathbb{R}^2$ taking (x, y) to (ax, ay) is invertible? Answer this two ways:

a. Give a formula for an inverse function $T_a^{-1} \colon \mathbb{R}^2 \to \mathbb{R}^2$ for T_a and verify that it really is the inverse of T_a.

b. Verify that T_a is one-to-one and onto.

So, we see that the term "symmetry" depends on context. It matters what features of the domain we pay attention to and seek to preserve, or leave the same (e.g., a geometric shape in a fixed position), and it matters what we mean by "the same." Put differently, it matters what attributes of the domain of the function we deem important. As we shall see, those attributes need not necessarily be geometric.

This text is about "symmetries" of mathematical systems* — an example will soon illustrate what we mean. As we go along, we will change our point of view from time to time, focusing on different mathematical systems as they suit our purposes. When he began to develop the ideas that bear his name, Sophus Lie (he was Norwegian, and his last name is pronounced "Lee") was actually interested in symmetries of *differential equations*, if you can imagine that!

We begin with symmetries of the vector space $\mathbb{R}^n$. (In the next section, we'll get more concrete by specializing to the plane $\mathbb{R}^2$.) To define symmetry in this setting, we need to decide what is important to "preserve." To begin, we seek to preserve the *vector space* structure of $\mathbb{R}^n$. What makes $\mathbb{R}^n$ a vector space? It is the existence of two operations, vector addition and scalar multiplication, satisfying certain axioms. Consequently, we define a *vector space symmetry* as an invertible function $T\colon \mathbb{R}^n \to \mathbb{R}^n$ that "preserves," or we could say "is compatible with," these operations in the following sense.

- It doesn't matter whether you first add two vectors and then apply T to the sum or first apply T to each vector and then add the results.
- Nor does it matter whether you first multiply a vector by a scalar and then apply T to the result or first apply the function to the vector and then multiply the result by the scalar.

In symbols, we say an invertible function $T\colon \mathbb{R}^n \to \mathbb{R}^n$ is a **vector space symmetry** of $\mathbb{R}^n$ if for all $\boldsymbol{v}, \boldsymbol{w} \in \mathbb{R}^n$ and all $a \in \mathbb{R}$

$$T(\boldsymbol{v} + \boldsymbol{w}) = T(\boldsymbol{v}) + T(\boldsymbol{w})$$
$$T(a\boldsymbol{v}) = aT(\boldsymbol{v}).$$

Do you recognize that a "vector space symmetry" is thus the same thing as an invertible linear transformation?

Problem 1.1.10. (S)

Every linear transformation is not only compatible with the operations of vector addition and scalar multiplication but also maps the zero vector to itself. Verify that if $T\colon \mathbb{R}^n \to \mathbb{R}^n$ is a linear transformation (not necessarily invertible), then $T(\mathbf{0}) = \mathbf{0}$, where $\mathbf{0} = (0, 0, \ldots, 0)$. Although you could make this argument using matrices, do it a different way: make use of the definition of a linear transformation and the fact that $\mathbf{0} + \mathbf{0} = \mathbf{0}$.

1.2 Distance is fundamental

We start in familiar territory, with the vector space $\mathbb{R}^2$, but from the point of view of the classic text *Spacetime Physics* by Taylor and Wheeler. *Spacetime Physics* is an introduction to special relativity. It opens with "The Parable of the Surveyors," a story

*Mathematicians use the term *morphism* for a function, not necessarily invertible, which preserves attributes of a particular mathematical structure.

about a town with two rival teams of surveyors, the Nighttime Surveyors and the Daytime Surveyors. Both teams use the center of the town square as the origin of a coordinate system for $\mathbb{R}^2$ that provides a map of the town. But the Nighttime Surveyors use the north star to determine their y-axis, and their rivals use magnetic north. As a consequence, the rivals assign different coordinates to the points of the town (except the center of the town square). One fine day (and night), an enterprising student who studies the two sets of coordinates discovers that, despite the disagreement, both teams agree on the *distance* between any two points in town. So, distance is something intrinsic to the points of the town, more fundamental than coordinates. Motivated by this example, we add distance to our list of what is important about $\mathbb{R}^2$ and so should be preserved by a symmetry. We thus study functions $T\colon \mathbb{R}^2 \to \mathbb{R}^2$ with the following properties.

- T is invertible.
- T is a linear transformation.
- For any points P and Q in $\mathbb{R}^2$, the distance from $T(P)$ to $T(Q)$ is the same as the distance from P to Q.

Such a linear transformation is called an **isometry** of $\mathbb{R}^2$ — the etymology is "same measurement," that is, same distance. Assume these properties hold for Problems 1.2.1–1.2.9; the goal of these problems is a very explicit description of these isometries.

Problem 1.2.1.
Suppose the distance from some point P in $\mathbb{R}^2$ to the origin $(0, 0)$ is $d > 0$. Explain why $T(P)$ must lie on a circle of radius d centered at the origin. (Don't do this using coordinates; reason geometrically, and use results from the previous section if needed.)

Problem 1.2.2.
Use the result of (1.2.1) to explain why $T(1, 0)$ and $T(0, 1)$ must each lie on the circle of radius 1 centered at the origin.

Problem 1.2.3.

a. Explain why you may assume $T(1, 0) = (\cos(\alpha), \sin(\alpha))$ for some real number α with $0 \le \alpha < 2\pi$. Draw a unit circle and mark a point on the unit circle in the first quadrant corresponding to some otherwise arbitrary α with $0 < \alpha < \pi/2$.

b. Use the assumptions about T and the result of (1.2.2) to explain why there are just two possible locations for $T(0, 1)$. Mark these two points on your figure from (a), and determine the coordinates of each of these points in terms of the angle α. (A listing of trigonometric identities is provided at the end of this section. Use them as needed to simplify the coordinates.)

As usual, the standard basis for the vector space $\mathbb{R}^2$ consists of $(1, 0)$ and $(0, 1)$. Our convention is that we write functions on the left and inputs on the right. So if a linear transformation T has matrix M with respect to the standard basis, and a vector $\boldsymbol{v}$ is written as a column with respect to the standard basis, then $T(\boldsymbol{v}) = M\boldsymbol{v}$, where the product on the right is matrix multiplication. As a consequence, the matrix of a transformation T with respect to the standard basis has for its columns the images under T of the standard basis vectors.

Problem 1.2.4.

Each choice for $T(0,1)$ in (1.2.3) determines a matrix for T. Write the two matrices, explaining which is which. You should get the two matrices:

$$\begin{bmatrix} \cos(\alpha) & -\sin(\alpha) \\ \sin(\alpha) & \cos(\alpha) \end{bmatrix} \text{ and } \begin{bmatrix} \cos(\alpha) & \sin(\alpha) \\ \sin(\alpha) & -\cos(\alpha) \end{bmatrix}.$$

Problem 1.2.5. (S)

For one of the two matrices in (1.2.4), T corresponds to a *rotation* of the plane counter-clockwise about the origin. Which one? (Look at your picture in (1.2.3).) What is the angle of the rotation? What is the *determinant* of this matrix?

Problem 1.2.6.

For the other matrix in (1.2.4), T corresponds to a *reflection* of the plane across a line through the origin, the *mirror line* of the reflection. What angle does this mirror line make with the positive x-axis? What is the determinant of this matrix?

The results of (1.2.3)–(1.2.6) remain valid no matter which quadrant $T(1,0)$ is in. (Can you show this?) Consequently, you may assume that the two matrices in (1.2.4) always correctly describe the two possibilities for T.

Problem 1.2.7.

Assume that T and S are both rotations of the plane about the origin, the matrix of T given in terms of the angle α as in (1.2.5), and the matrix of S given similarly, but in terms of an angle β. Use matrix multiplication to find the matrix corresponding to the composition $T \circ S$. Simplify the matrix entries using trigonometric identities, and use the simplified matrix to show that $T \circ S$ is also a rotation. What is the angle of this rotation? Does your answer make geometric sense? What can you say about the composition in the other order, $S \circ T$?

Problem 1.2.8. (S)

Now consider the composition $T \circ S$ of a rotation T (expressed in terms of an angle α as in (1.2.5)) and a reflection S (expressed in terms of an angle β, as in (1.2.6)). Describe the result as a matrix and geometrically. Does the order of the composition matter? Illustrate with pictures for the case when the angle of rotation is $\pi/2$ (counter-clockwise) and the mirror line for the reflection is $y = x$.

Problem 1.2.9.

Finally, repeat (1.2.8) assuming that T and S are both reflections. Illustrate with pictures for the case when the mirror lines for T and S are $y = x$ and the y-axis respectively.

Problem 1.2.10. (C)

Actually, it was unnecessary to assume that T was an invertible linear transformation in the beginning of this section. Assume only that $T\colon \mathbb{R}^2 \to \mathbb{R}^2$, $T(0,0) = (0,0)$, and T preserves distance.

a. Using pictures, give an informal geometric argument that T is a linear transformation by interpreting vectors as directed line-segments of the form OP, where $O = (0, 0)$, and using geometric descriptions of vector addition (the "parallelogram rule") and scalar multiplication.

b. Assuming the result of (a), use facts from linear algebra to explain why T must be invertible.

Trignometric Identities

$$\cos(0) = \sin(\pi/2) = 1$$

$$\sin(0) = \cos(\pi/2) = 0$$

$$\cos(-\alpha) = \cos(\alpha) \qquad \sin(-\alpha) = -\sin(\alpha)$$

$$\sin(\alpha + \beta) = \sin(\alpha)\cos(\beta) + \cos(\alpha)\sin(\beta)$$

$$\cos(\alpha + \beta) = \cos(\alpha)\cos(\beta) - \sin(\alpha)\sin(\beta)$$

$$\sin(\alpha - \beta) = \sin(\alpha)\cos(\beta) - \cos(\alpha)\sin(\beta)$$

$$\cos(\alpha - \beta) = \cos(\alpha)\cos(\beta) + \sin(\alpha)\sin(\beta)$$

1.3 Groups of symmetries

This section introduces the definition of a *group* of symmetries and also extends the two languages we will use almost interchangeably: the language of matrices and the language of linear transformations. It also connects the ideas of *preserving distance* and *preserving the dot product.* Using the language of dot products and matrices will help us generalize from $\mathbb{R}^2$ or $\mathbb{R}^3$ to $\mathbb{R}^n$ for any positive integer n. It will also open the door to new geometries, which we'll explore in section 4.

A set G of linear transformations from a vector space to itself forms a **group under composition of functions** provided the following conditions hold.

- The composition of two transformations in G is again a transformation in G. In other words, if S and T belong to G then so does the composition $S \circ T$. We say G is **closed** under composition.
- The **identity** transformation I is in G, where I is defined by $I(\boldsymbol{v}) = \boldsymbol{v}$ for all vectors $\boldsymbol{v}$ in the vector space.
- If a transformation T is in G, then it is invertible, and its **inverse** T^{-1} is also in G, where $T \circ T^{-1} = T^{-1} \circ T = I$.

A group operation is required to obey the **associative law**:

$$T \circ (R \circ S) = (T \circ R) \circ S$$

for all T, R, S in G. Composition of functions is always associative. Also, we know from linear algebra that the composition of two linear transformations is always a linear transformation and that the inverse of an invertible linear transformation is an invertible linear transformation.

Equivalently, for a fixed positive integer n, a set G of n by n matrices forms a **group under matrix multiplication** provided the following conditions hold.

- The product of two matrices in G is again a matrix in G. In other words, if M and N belong to G then so does the product MN. We say G is **closed** under matrix multiplication.
- The n by n **identity** matrix I_n is in G.
- If a matrix M is in G, then it is invertible and its **inverse** M^{-1} is also in G. (Recall $MM^{-1} = M^{-1}M = I_n$.)

Matrix multiplication always satisfies the **associative law**:

$$M(NQ) = (MN)Q$$

for all n by n matrices M, N, Q. Also, we know that the product of two n by n matrices is an n by n matrix, and the inverse of an invertible n by n matrix is an invertible n by n matrix.

Whether talking about transformations or matrices, a **subgroup** of a group G is a nonempty subset of G that is itself a group under the same operation as G.

A linear transformation $T: \mathbb{R}^n \to \mathbb{R}^n$ corresponds to an n by n matrix M, its matrix with respect to the standard basis of $\mathbb{R}^n$: $T(\boldsymbol{v}) = M\boldsymbol{v}$, for every column vector $\boldsymbol{v}$ in $\mathbb{R}^n$. In the same way, an n by n matrix M corresponds to a linear transformation $T: \mathbb{R}^n \to \mathbb{R}^n$. Further, if the transformations S and T correspond to the matrices N and M respectively, then the composition of linear transformations $S \circ T$ corresponds to multiplication of the corresponding matrices NM.

We will sometimes use the language of linear transformations and sometimes the language of matrices, whichever is more convenient or seems more illuminating. Generally, what we can say in one language can also be said in the other.

For any choice of the positive integer n, we will denote the set of *all* n by n matrices with real number entries by $\mathcal{M}(n, \mathbb{R})$. The set of all *invertible* n by n matrices is denoted $GL(n, \mathbb{R})$, and is called the **general linear group** of degree n over the real numbers,

$$GL(n, \mathbb{R}) = \{M \in \mathcal{M}(n, \mathbb{R}) \ : \ \det(M) \neq 0\}.$$

(Recall that a square matrix M is invertible if and only if $\det(M) \neq 0$.) Correspondingly, the set of all *invertible* linear transformations $\mathbb{R}^n \to \mathbb{R}^n$ is denoted by

$$GL(\mathbb{R}^n) = \{T: \mathbb{R}^n \to \mathbb{R}^n \ : \ T \text{ linear and invertible }\}.$$

Recall that a linear transformation $\mathbb{R}^n \to \mathbb{R}^n$ is invertible if and only if the determinant of its corresponding matrix is nonzero. In fact, since every matrix representing T has the *same* determinant, we can unambiguously speak of the determinant of T. So, T is invertible if and only if $\det(T) \neq 0$.

In the following problem you'll verify that $GL(n, \mathbb{R})$ really is a group under matrix multiplication, and $GL(\mathbb{R}^n)$ really is a group under composition of functions. The group $GL(\mathbb{R}^n)$ is also called the **general linear group** in dimension n. In a sense which we'll make precise in Chapter 3, these two groups are essentially *the same*. (Using formal language, we say they are *isomorphic* groups.) That justifies using the same name for both groups. We use slightly different notation for the two groups as an aid to keeping straight whether we are talking about matrices or linear transformations.

The general linear group, in either form, is our "mother group." The set of n by n matrices with determinant 1 is also a group under matrix multiplication. This is our first "daughter group." It is denoted $SL(n, \mathbb{R})$ and is called the **special linear group** of degree n over the real numbers,

$$SL(n, \mathbb{R}) = \{M \in \mathcal{M}(n, \mathbb{R}) : \det(M) = 1\}.$$

More formally, $SL(n, \mathbb{R})$ is a *subgroup* of $GL(n, \mathbb{R})$. Similarly, $SL(\mathbb{R}^n)$ is a subgroup of $GL(\mathbb{R}^n)$. The reason we call the general linear group the "mother group" is that almost all the groups we study in this text will be subgroups of the general linear group.

Problem 1.3.1.

a. Verify that $GL(n, \mathbb{R})$ is a group under matrix multiplication. (Recall $\det(MN) = \det(M)\det(N)$ and $\det(I_n) = 1$.)

b. Verify that $SL(n, \mathbb{R})$ is a group under matrix multiplication.

c. Verify that $GL(\mathbb{R}^n)$ is a group under composition of functions.

Problem 1.3.2. (S)

a. Show that the set of all *rotations* (which you should think of as the set of all 2 by 2 matrices having the form specified in (1.2.5)) is a group under matrix multiplication. Don't re-prove previous results; just cite them as needed.

b. Is the set of all reflections (the matrices of (1.2.6)) a group under matrix multiplication? Explain.

The group of (1.3.2a) is called the **special orthogonal group** of degree 2 over the real numbers and is denoted by $SO(2, \mathbb{R})$,

$$SO(2, \mathbb{R}) = \left\{ \begin{bmatrix} \cos(\alpha) & -\sin(\alpha) \\ \sin(\alpha) & \cos(\alpha) \end{bmatrix} : \alpha \in \mathbb{R} \right\}.$$

Problem 1.3.3.

Show that the set of all *rotations and reflections* (which you should think of as the set of all 2 by 2 matrices having the forms in (1.2.4)) is a group under matrix multiplication. Don't re-prove previous results; cite them as needed.

The group in (1.3.3) is called the **orthogonal group** of degree 2 over the real numbers and is denoted by $O(2, \mathbb{R})$.

$$O(2, \mathbb{R}) = \left\{ \begin{bmatrix} \cos(\alpha) & -\sin(\alpha) \\ \sin(\alpha) & \cos(\alpha) \end{bmatrix}, \begin{bmatrix} \cos(\alpha) & \sin(\alpha) \\ \sin(\alpha) & -\cos(\alpha) \end{bmatrix} : \alpha \in \mathbb{R} \right\}.$$

The group $SO(2, \mathbb{R})$ of (1.3.2) is a *subgroup* of $O(2, \mathbb{R})$.

Now we return to linear transformations. The linear transformation versions of the groups in (1.3.2a) and (1.3.3) are denoted by $SO(\mathbb{R}^2)$ and $O(\mathbb{R}^2)$ respectively.

The next problem relates the notion of *preserving distance* to *preserving the dot product*. Recall the dot product on $\mathbb{R}^n$:

$$\boldsymbol{v} \cdot \boldsymbol{w} = \boldsymbol{v}^T \boldsymbol{w} = v_1 w_1 + v_2 w_2 + \cdots + v_n w_n,$$

for $\boldsymbol{v} = (v_1, v_2, \ldots, v_n)$ and $\boldsymbol{w} = (w_1, w_2, \ldots, w_n)$. Because we are writing functions on the left, we usually write vectors as columns. The notation $\boldsymbol{v}^T$ means the transpose of $\boldsymbol{v}$ and is a row vector. We say that a linear transformation $T\colon \mathbb{R}^n \to \mathbb{R}^n$ **preserves the dot product** if $T(\boldsymbol{v}) \cdot T(\boldsymbol{w}) = \boldsymbol{v} \cdot \boldsymbol{w}$ for all vectors $\boldsymbol{v}, \boldsymbol{w}$ in $\mathbb{R}^n$.

Although the following problem is for $\mathbb{R}^3$, the arguments are the same for $\mathbb{R}^n$ for any positive integer n. Assume $O = (0, 0, 0)$ is the origin of $\mathbb{R}^3$. For any points P and Q in $\mathbb{R}^3$, define vectors $\boldsymbol{p}$ and $\boldsymbol{q}$ in $\mathbb{R}^3$ by $\boldsymbol{p} = OP$ (meaning the directed line segment from O to P) and $\boldsymbol{q} = OQ$. So the coordinates of the point $P = (p_1, p_2, p_3)$ and the entries in the vector $\boldsymbol{p} = (p_1, p_2, p_3)$ are the same. For any linear transformation $T\colon \mathbb{R}^3 \to \mathbb{R}^3$, $T(O) = O$ by (1.1.10); it follows that $T(\boldsymbol{p}) = OT(P)$ and $T(\boldsymbol{q}) = OT(Q)$. This notation permits us to switch back and forth between thinking of the elements of $\mathbb{R}^3$ as *points* or as *vectors*. Note, however, that the defining properties of a linear transformation (e.g., $T(\boldsymbol{p} - \boldsymbol{q}) = T(\boldsymbol{p}) - T(\boldsymbol{q})$) only make sense for vectors, since we can't add or subtract points.

Problem 1.3.4.

Use the preceding notation for this problem.

a. Show that $\sqrt{\boldsymbol{p} \cdot \boldsymbol{p}}$ is the usual Euclidean length of the vector $\boldsymbol{p} = (p_1, p_2, p_3)$ in $\mathbb{R}^3$ thought of as the directed line segment OP.

b. Show that $\sqrt{(\boldsymbol{p} - \boldsymbol{q}) \cdot (\boldsymbol{p} - \boldsymbol{q})}$ is the distance from P to Q. For a linear transformation $T\colon \mathbb{R}^3 \to \mathbb{R}^3$, give a similar description of the distance from $T(P)$ to $T(Q)$ in terms of $T(\boldsymbol{p} - \boldsymbol{q})$.

c. Assume that $T\colon \mathbb{R}^3 \to \mathbb{R}^3$ is a linear transformation that *preserves the dot product.* Show that T must *preserve distances*; that is, the distance between P and Q is the same as the distance between $T(P)$ and $T(Q)$ for any points P and Q in $\mathbb{R}^3$.

d. Now assume $T\colon \mathbb{R}^3 \to \mathbb{R}^3$ is a linear transformation that *preserves distances* in $\mathbb{R}^3$. Show that T must *preserve the dot product.*

Problem (1.3.4) shows that the set of linear transformations of $\mathbb{R}^3$ that preserve distance is the same as the set that preserve the dot product. This set is called the (Euclidean) **orthogonal group** in three dimensions, and it is denoted $O(\mathbb{R}^3)$.

$$\begin{aligned} O(\mathbb{R}^3) &= \{T \in GL(\mathbb{R}^3) \,:\, T \text{ preserves distance}\} \\ &= \{T \in GL(\mathbb{R}^3) \,:\, T \text{ preserves the dot product}\}. \end{aligned}$$

The next problem will show that $O(\mathbb{R}^3)$ really is a group under composition of functions, and the subset of transformations having determinant $+1$ is a subgroup. By analogy to the 2-dimensional case, this subset is called the **rotation group** or the **special orthogonal group** of degree 3 over the real numbers, and it is denoted by $SO(\mathbb{R}^3)$,

$$SO(\mathbb{R}^3) = \{T \in O(\mathbb{R}^3) \,:\, \det(T) = +1\}.$$

We'll see in Chapter 2 that the name "rotation group" is more than just an analogy.

Problem 1.3.5. (S)

Show that

$$O(\mathbb{R}^3) = \{T \in GL(\mathbb{R}^3) \,:\, T(\boldsymbol{v}) \cdot T(\boldsymbol{w}) = \boldsymbol{v} \cdot \boldsymbol{w} \text{ for all } \boldsymbol{v}, \boldsymbol{w} \in \mathbb{R}^3\}$$

is a group under composition of functions by carrying out the following steps.

a. Show that $O(\mathbb{R}^3)$ contains the *identity* transformation I.

b. Assume S and T are in $O(\mathbb{R}^3)$. Show that $T \circ S$ preserves the dot product. Since we know from linear algebra that the composition of linear transformations is linear, we conclude that $T \circ S$ is in $O(\mathbb{R}^3)$, so $O(\mathbb{R}^3)$ is *closed*.

c. Show that if $\boldsymbol{v}$ is a particular vector in $\mathbb{R}^3$ satisfying $\boldsymbol{v} \cdot \boldsymbol{w} = 0$ for *all* $\boldsymbol{w}$ in $\mathbb{R}^3$, then $\boldsymbol{v} = (0, 0, 0)$.

d. Assume $T \in O(\mathbb{R}^3)$. Use (c) to show that if $\boldsymbol{v}$ is a particular vector in $\mathbb{R}^3$ satisfying $T(\boldsymbol{v}) = (0, 0, 0)$, then $\boldsymbol{v} = (0, 0, 0)$.

e. Result (d) implies that if T is in $O(\mathbb{R}^3)$, then the kernel of T consists of the zero vector only, so T is invertible; write T^{-1} to denote its *inverse*. Show that T^{-1} preserves the dot product. Since we know from linear algebra that the inverse of an invertible linear transformation is a linear transformation, we conclude that T^{-1} is in $O(\mathbb{R}^3)$. This completes the proof that $O(\mathbb{R}^3)$ is a group.

As by now you may be expecting, there is a matrix group that corresponds to $O(\mathbb{R}^3)$. Defining it requires some preliminary steps describing how to recognize the matrix of a linear transformation that preserves the dot product. The details are in the next problem.

Problem 1.3.6.

Suppose the linear transformation $S\colon \mathbb{R}^3 \to \mathbb{R}^3$ has matrix M with respect to the standard basis of $\mathbb{R}^3$, so $S(\boldsymbol{v}) = M\boldsymbol{v}$, where $\boldsymbol{v}$ is written as a column vector. Recall that $\boldsymbol{v} \cdot \boldsymbol{w} = \boldsymbol{v}^T \boldsymbol{w}$, where $\boldsymbol{v}, \boldsymbol{w} \in \mathbb{R}^3$ are written as column vectors, so $\boldsymbol{v}^T$ is a row vector.

a. Show that $S(\boldsymbol{v}) \cdot S(\boldsymbol{w}) = \boldsymbol{v}^T(M^T M)\boldsymbol{w}$.

b. Explain why (a) shows that S preserves the dot product on $\mathbb{R}^3$ if and only if $\boldsymbol{v}^T(M^T M)\boldsymbol{w} = \boldsymbol{v}^T\boldsymbol{w}$ for all $\boldsymbol{v}, \boldsymbol{w} \in \mathbb{R}^3$.

c. Let $A \in \mathcal{M}(3, \mathbb{R})$. Show that $\boldsymbol{v}^T A\boldsymbol{w} = \boldsymbol{v}^T\boldsymbol{w}$ for all $\boldsymbol{v}, \boldsymbol{w} \in \mathbb{R}^3$ if and only if $A = I_3$. [Hint: choose $\boldsymbol{v} = \boldsymbol{e}_i$ and $\boldsymbol{w} = \boldsymbol{e}_j$, where $\boldsymbol{e}_i$ is the standard basis vector consisting of all zeroes except a 1 in the ith coordinate. What is $\boldsymbol{e}_i^T A\boldsymbol{e}_j$ in terms of the entries of A? You don't need to do nine specific calculations; what happens in general?]

d. Show that if the linear transformation $S\colon \mathbb{R}^3 \to \mathbb{R}^3$ has matrix M, then S is in $O(\mathbb{R}^3)$ if and only if $M^T M = I_3$.

Using the result of (1.3.6d), we write

$$O(3, \mathbb{R}) = \{M \in \mathcal{M}(3, \mathbb{R}) \;:\; M^T M = I_3\}.$$

As is true for the two forms of the general linear group, the groups $O(3, \mathbb{R})$ and $O(\mathbb{R}^3)$ are essentially *the same*— i.e., they are isomorphic.

We have a simple test for determining whether a 3 by 3 matrix M is in $O(3, \mathbb{R})$: if $M^T M = I_3$, it is, and otherwise it isn't. Similarly, if we have been told to assume that a matrix M is in $O(3, \mathbb{R})$ as part of a hypothesis, then we know we can assume it passes the test: $M^T M = I_3$.

Problem 1.3.7. (S)

Show that $O(3, \mathbb{R})$ is a group under matrix multiplication by carrying out the following steps.

a. Show that I_3 is in $O(3, \mathbb{R})$.
b. Show that if M_1 and M_2 are in $O(3, \mathbb{R})$, then $M_1 M_2$ is in $O(3, \mathbb{R})$.
c. Show that if M is in $O(3, \mathbb{R})$, then M is invertible and M^{-1} is in $O(3, \mathbb{R})$.
d. This part isn't necessary for the proof that $O(3, \mathbb{R})$ is a group, but it will turn out to have interesting geometric consequences, which we'll pursue in Chapter 2. Show that $M \in O(3, \mathbb{R})$ implies $\det M = \pm 1$.

Matrix language makes it natural to generalize the definition of the orthogonal group to higher dimensions. For any positive integer n, define the **orthogonal group of degree n over the real numbers** by

$$O(n, \mathbb{R}) = \{M \in \mathcal{M}(n, \mathbb{R}) : M^T M = I_n\}.$$

Similarly,

$$\begin{aligned} SO(n, \mathbb{R}) &= \{M \in O(n, \mathbb{R}) : \det M = 1\} \\ &= \{M \in \mathcal{M}(n, \mathbb{R}) : M^T M = I_3 \text{ and } \det M = 1\} \end{aligned}$$

is the **special orthogonal group of degree n over the real numbers**.

Problem 1.3.8.

a. Show that $O(n, \mathbb{R})$ is a group under matrix multiplication.
b. Show that $SO(n, \mathbb{R})$ is a subgroup of $O(n, \mathbb{R})$.

Problem 1.3.9.

This problem is a small digression on *finite* (discrete) symmetry groups. Return to two dimensions and imagine a paper square placed on the coordinate axes with its center at the origin and its sides parallel to the axes, as in section 1. This square has eight symmetries, four rotations (one of which is the identity) and four reflections. Regard these symmetries as elements of $O(2, \mathbb{R})$.

a. Write the four matrices of these rotations.
b. Write the four matrices of these reflections.
c. Show that the four matrices in (a) form a group under matrix multiplication. Organize your work by first filling out a *multiplication table* for the four matrices. In other words, if $M_1 = I_2$, M_2, M_3, M_4 are the four matrices in (a), fill in the table below, where $\times$ denotes matrix multiplication. For example, the second entry in the third row should be the result of the multiplication $M_3 M_2$.

$\times$	M_1	M_2	M_3	M_4
M_1				
M_2				
M_3				
M_4				

d. Show that the 8 matrices in (a) and (b) form a group under matrix multiplication. Give names to the four reflection matrices and organize your work by filling out an 8 by 8 multiplication table for the 8 matrices.

Remark Problem (1.3.9) shows that even though $O(2, \mathbb{R})$ is an infinite group, it contains finite subgroups. It turns out that it is possible to give a complete description of *all* the finite subgroups of $O(2, \mathbb{R})$. (See for example *Groups and Symmetry: A Guide to Discovering Mathematics* by Farmer.)

Problem 1.3.10.

Explain why every transformation in $O(\mathbb{R}^2)$ may be regarded as a symmetry of the paper disk in (1.1.6).

Problem 1.3.11. (C)

This problem involves a further generalization of the orthogonal group. Let n be a positive integer, and fix a matrix $C \in \mathcal{M}(n, \mathbb{R})$. Let

$$G = \{M \in \mathcal{M}(n, \mathbb{R}) \ : \ M^T C M = C\}.$$

When $C = I_n$, G is the (Euclidean) orthogonal group, the matrices corresponding to the transformations that preserve the standard (Euclidean) dot product on $\mathbb{R}^n$. But other choices of C are possible.

a. What conditions on C guarantee that G is a group? For C satisfying your conditions, show that G is a group under matrix multiplication.

b. Are your conditions necessary as well as sufficient? Support your answer with an argument or a counter example.

c. Suppose that the matrix C is chosen so that G is a group. Fix $r \in \mathbb{R}$, $r \neq 0$, and let $C' = rC$. Let $G' = \{M \in \mathcal{M}(n, \mathbb{R}) \ : \ M^T C' M = C'\}$. Show that $G' = G$.

1.4 Bilinear forms and symmetries of spacetime

The parable of the surveyors in Taylor and Wheeler's *Spacetime Physics* sets up an analogy for their introduction to the theory of special relativity. Instead of the town, they consider the universe, spacetime. For simplicity, they first assume the universe has just one spatial dimension, and they describe it as if it were an infinitely long, infinitely skinny "one-dimensional" tunnel along an x-axis. Instead of rival teams of surveyors, now there are rival laboratories, each observing events in the universe.

One (tiny!) laboratory is stationary, located at $x = 0$. The other is a rocket moving with constant velocity through the tunnel. When an *event* occurs, an observer in each laboratory makes measurements to determine the event's spatial coordinate x and its time coordinate t. Both observers agree on the coordinates of a reference event at the location $x = 0$ and the time $t = 0$, but otherwise they assign different coordinates to the same events. However, by analogy to distance in the parable of the surveyors, both observers agree on what physicists call the *interval* between two events. If an observer assigns the coordinates $\boldsymbol{p} = (t, x)$ and $\boldsymbol{q} = (t', x')$ to two events P and Q in this simplified 2-dimensional version of spacetime (and if time is measured in suitable units), the interval between P and Q satisfies

$$(\text{interval})^2 = (t' - t)^2 - (x' - x)^2.$$

(See page 24 of Taylor and Wheeler, or page 6 in the second edition.) It would be more natural to write an equation for the interval itself, but — since the difference of two squares can be negative — in order to handle all cases, we'd have to work over the complex numbers. For now, we'll stick to real numbers.

Motivated by this newly important quantity, we define a new "dot product" we call the **Lorentz product** on $\mathbb{R}^2$ by

$$(t_1, x_1) \cdot (t_2, x_2) = t_1 t_2 - x_1 x_2 .$$

(Hendrik Lorentz was a Dutch physicist who, independently of Einstein, contributed significantly to the theory of special relativity. The interval is sometimes called the *Lorentz interval*, which motivates our use of Lorentz's name for the dot product.) Using this new dot product, we can write

$$(\text{interval})^2 = (\boldsymbol{q} - \boldsymbol{p}) \cdot (\boldsymbol{q} - \boldsymbol{p}).$$

In this section, when we write $\boldsymbol{v} \cdot \boldsymbol{w}$ for vectors $\boldsymbol{v}$ and $\boldsymbol{w}$ in $\mathbb{R}^2$, we mean the Lorentz product of the vectors.

Problem 1.4.1.

Check that the Lorentz product has the following properties.

a. $\boldsymbol{v} \cdot \boldsymbol{w} = \boldsymbol{w} \cdot \boldsymbol{v}$ for all $\boldsymbol{v}, \boldsymbol{w}$ in $\mathbb{R}^2$.

b. For all $\boldsymbol{u}, \boldsymbol{v}, \boldsymbol{w}$ in $\mathbb{R}^2$,

$$\begin{aligned} (\boldsymbol{u} + \boldsymbol{v}) \cdot \boldsymbol{w} &= \boldsymbol{u} \cdot \boldsymbol{w} + \boldsymbol{v} \cdot \boldsymbol{w}, \\ \boldsymbol{u} \cdot (\boldsymbol{v} + \boldsymbol{w}) &= \boldsymbol{u} \cdot \boldsymbol{v} + \boldsymbol{u} \cdot \boldsymbol{w}. \end{aligned}$$

c. $(a\boldsymbol{v}) \cdot \boldsymbol{w} = a(\boldsymbol{v} \cdot \boldsymbol{w}) = \boldsymbol{v} \cdot (a\boldsymbol{w})$ for all $\boldsymbol{v}, \boldsymbol{w}$ in $\mathbb{R}^2$ and all a in $\mathbb{R}$.

d. **(S)** If $\boldsymbol{v}$ is a particular vector satisfying

$$\boldsymbol{v} \cdot \boldsymbol{w} = 0 \quad \text{for } \underline{\text{every}} \quad \boldsymbol{w} \quad \text{in} \quad \mathbb{R}^2,$$

then $\boldsymbol{v}$ must be the zero vector $(0, 0)$.

Formal language Property (a) means the Lorentz product is **symmetric**; properties (b) and (c) mean it is **bilinear**; property (d) means it is **non-degenerate**. The usual dot product on $\mathbb{R}^n$ also has these properties. (The argument that the usual dot product on $\mathbb{R}^n$ is non-degenerate is essentially the same as (1.3.5c).) The more general term encompassing all such "dot products" is a symmetric, non-degenerate bilinear **form**. A vector space together with a symmetric, non-degenerate bilinear form is called a **metric vector space**. Linear transformations of the vector space that preserve the bilinear form are called **isometries** of the metric vector space. Note that, as in (1.3.5e), if a linear transformation preserves a *non-degenerate* bilinear form, it can be shown that it has to be invertible. You'll make the argument again in (1.4.3d).

Problem 1.4.2.

The set $\mathcal{M}(n, \mathbb{R})$ is a vector space, and it's possible to define a form on this space by $A \cdot B = \operatorname{tr}(AB)$ for $A, B \in \mathcal{M}(n, \mathbb{R})$.

a. Show that this form is symmetric.

b. Show that this form is bilinear.

c. Is the form non-degenerate? Justify your answer.

Returning to the Lorentz product, in the next problem you'll use the language of transformations to show that

$$\mathcal{L}(\mathbb{R}^2) = \{T\colon \mathbb{R}^2 \to \mathbb{R}^2 \ :\ T \text{ linear and } T(\boldsymbol{v}) \cdot T(\boldsymbol{w}) = \boldsymbol{v} \cdot \boldsymbol{w} \text{ for all } \boldsymbol{v}, \boldsymbol{w} \in \mathbb{R}^2\}$$

is a group under composition of functions. It is called the **Lorentz group**. The Lorentz group has a subgroup $\mathcal{L}^+(\mathbb{R}^2)$ consisting of its elements of determinant 1 and called the group of **Lorentz rotations**,

$$\mathcal{L}^+(\mathbb{R}^2) = \{T \in \mathcal{L}(\mathbb{R}^2) \ :\ \det(T) = 1\ \}.$$

Problem 1.4.3.

Show that $\mathcal{L}(\mathbb{R}^2)$ is a group of linear transformations by carrying out steps (a)–(d) below.

a. Show that $\mathcal{L}(\mathbb{R}^2)$ contains the *identity* transformation I.

b. Assume S and T are in $\mathcal{L}(\mathbb{R}^2)$. Show that $T \circ S$ preserves the Lorentz product. Since we know that the composition of linear transformations is linear, this shows that $T \circ S$ is in $\mathcal{L}(\mathbb{R}^2)$, so $\mathcal{L}(\mathbb{R}^2)$ is *closed.*

c. Assume T is in $\mathcal{L}(\mathbb{R}^2)$. Show that if $\boldsymbol{v}$ is a particular vector satisfying $T(\boldsymbol{v}) = (0, 0)$, then $\boldsymbol{v} = (0, 0)$.

d. Result (c) implies that if T is in $\mathcal{L}(\mathbb{R}^2)$, then T is invertible; write T^{-1} to denote its *inverse.* Show that T^{-1} preserves the Lorentz product. Since we know that the inverse of an invertible linear transformation is a linear transformation, this shows that T^{-1} is in $\mathcal{L}(\mathbb{R}^2)$ and completes the proof that $\mathcal{L}(\mathbb{R}^2)$ is a group.

e. Show that $\mathcal{L}^+(\mathbb{R}^2)$ is a group; that is, $\mathcal{L}^+(\mathbb{R}^2)$ is a *subgroup* of $\mathcal{L}(\mathbb{R}^2)$.

Problem 1.4.4.

Let $\boldsymbol{v}, \boldsymbol{w} \in \mathbb{R}^2$. Show that $\boldsymbol{v} \cdot \boldsymbol{w} = \boldsymbol{v}^T C \boldsymbol{w}$ for

$$C = \begin{bmatrix} 1 & 0 \\ 0 & -1 \end{bmatrix}.$$

Problem 1.4.5.

a. Explain why the matrix group corresponding to $\mathcal{L}(\mathbb{R}^2)$ is

$$\mathcal{L}(2, \mathbb{R}) = \{M \in \mathcal{M}(2, \mathbb{R}) \ :\ M^T C M = C\},$$

with C as in (1.4.4).

b. Show that $\mathcal{L}(2, \mathbb{R})$ is a group under matrix multiplication.

Taylor and Wheeler's definition of the interval (and our definitions of the Lorentz form and the Lorentz group for $\mathbb{R}^2$) are specializations of the definitions for $\mathbb{R}^4$. We ordinarily think of *spacetime* as 4-dimensional, with an event at time t and in location (x, y, z) in $\mathbb{R}^3$ having coordinates (t, x, y, z). In $\mathbb{R}^4$, the Lorentz product corresponds to $\boldsymbol{v} \cdot \boldsymbol{w} = \boldsymbol{v}^T C \boldsymbol{w}$ for

$$C = \begin{bmatrix} 1 & 0 & 0 & 0 \\ 0 & -1 & 0 & 0 \\ 0 & 0 & -1 & 0 \\ 0 & 0 & 0 & -1 \end{bmatrix}.$$

In other words, we define the Lorentz product of $\boldsymbol{v} = (t_1, x_1, y_1, z_1)$ and $\boldsymbol{w} = (t_2, x_2, y_2, z_2)$ by

$$\boldsymbol{v} \cdot \boldsymbol{w} = t_1 t_2 - x_1 x_2 - y_1 y_2 - z_1 z_2.$$

Then the square of the *interval* between the spacetime events with coordinates $\boldsymbol{v}$ and $\boldsymbol{w}$ is again $(\boldsymbol{w} - \boldsymbol{v}) \cdot (\boldsymbol{w} - \boldsymbol{v})$. It is straightforward to check that this product has all the properties in (1.4.1). In this case, the set of matrices $\mathcal{L}(4, \mathbb{R})$ (or, equivalently, the set of linear transformations $\mathcal{L}(\mathbb{R}^4)$) is also called the **Lorentz group**.

Remarks

- Some authors use a dot product with matrix $-C$ for spacetime. By (1.3.11c), the corresponding group of isometries, the Lorentz group, is the same whether the matrix of the form is C or $-C$.
- In addition to calling the interval the *Lorentz interval*, it is standard to refer to the isometries of spacetime as *Lorentz transformations*, and we have used these conventions to name the Lorentz product, which doesn't have a standard name in the literature. However, the four-dimensional spacetime geometry is the invention of the German mathematician Hermann Minkowski, who was inspired by the work of Lorentz and Einstein, and many authors refer to the spacetime geometry as Minkowski space. For example, Callahan calls this bilinear form the Minkowski inner product in his *The Geometry of Spacetime*.

The analogy between distance and the interval is deeper than just the relationship between the standard Euclidean dot product and the Lorentz product. Recall the specific descriptions of matrices in $O(2, \mathbb{R})$ from section 2; those matrices had entries involving the (circular) trigonometric functions sine and cosine. We will see that the matrices in the Lorentz group $\mathcal{L}(2, \mathbb{R})$ have analogous specific descriptions involving the **hyperbolic trigonometric** functions. The hyperbolic sine and hyperbolic cosine are defined by

$$\sinh(\alpha) = \frac{e^{\alpha} - e^{-\alpha}}{2} \quad \text{and} \quad \cosh(\alpha) = \frac{e^{\alpha} + e^{-\alpha}}{2}$$

for $\alpha \in \mathbb{R}$ (and $\tanh(\alpha) = \sinh(\alpha)/\cosh(\alpha)$, etc.).

Problem 1.4.6.

Assuming the usual properties of the exponential function, verify the following hyperbolic trigonometric identities:

a. $\cosh^2(\alpha) - \sinh^2(\alpha) = 1$.

b. $\cosh(-\alpha) = \cosh(\alpha)$ and $\sinh(-\alpha) = -\sinh(\alpha)$.

c. $\cosh(\alpha + \beta) = \cosh(\alpha)\cosh(\beta) + \sinh(\alpha)\sinh(\beta)$,
$\sinh(\alpha + \beta) = \sinh(\alpha)\cosh(\beta) + \cosh(\alpha)\sinh(\beta)$.

d. $\cosh(\alpha) > 0$ for all α.

e. $(\cosh(x))' = \sinh(x)$ and $(\sinh(x))' = \cosh(x)$. (Here we use the prime to denote the derivative with respect to x.)

The usual trigonometric functions are called *circular* because every point on the circle with equation $x^2 + y^2 = 1$ can be written in the form $(\cos(\alpha), \sin(\alpha))$ for some α in $\mathbb{R}$. These functions are called *hyperbolic* trigonometric functions because of the properties

(1.4.5a) and (1.4.5d); every point on the right-hand branch of the hyperbola $x^2 - y^2 = 1$ can be written in the form $(\cosh(\alpha), \sinh(\alpha))$ for some α in $\mathbb{R}$; every point on the left-hand branch can be written in the form $(-\cosh(\alpha), \sinh(\alpha))$. To see why these are called hyperbolic *trigonometric* functions, do (1.4.5f) below. See Taylor and Wheeler, page 59 of the first edition, or Callahan, pages 44–45, for more on the comparison of the circular and hyperbolic trigonometric functions.

f. Use the Taylor series centered at $x = 0$ for e^x

$$e^x = 1 + x + \frac{x^2}{2!} + \frac{x^3}{3!} + \cdots + \frac{x^m}{m!} + \cdots$$

to write the Taylor series for $\sinh(x)$ and for $\cosh(x)$. Compare these to the series for $\sin(x)$ and $\cos(x)$:

$$\sin(x) = x - \frac{x^3}{3!} + \frac{x^5}{5!} + \cdots + (-1)^m \frac{x^{2m+1}}{(2m+1)!} + \cdots,$$

$$\cos(x) = 1 - \frac{x^2}{2!} + \frac{x^4}{4!} + \cdots + (-1)^m \frac{x^{2m}}{(2m)!} + \cdots.$$

Problem 1.4.7. (S)

Suppose $T(1,0) = (a,b)$ and $T(0,1) = (c,d)$ for T in $\mathcal{L}(2,\mathbb{R})$. Assume $a > 0$.

a. Explain why $T(1,0) = (\cosh(\alpha), \sinh(\alpha))$ for some α in $\mathbb{R}$.

b. Use $(1,0)\cdot(0,1) = 0$ to express c in terms of α and d.

c. Use $(0,1)\cdot(0,1) = -1$ to find *two* solutions for (c,d) in terms of α.

d. Write the matrix of T corresponding to each solution in (c). Find the determinant of each.

Following the Euclidean language in this case, we'll call the transformation of (1.4.7d) of determinant $+1$ a **rotation of the first kind** and the one of determinant -1 a **reflection of the first kind**.

Problem 1.4.8. (S)

Suppose T and S are rotations as in (1.4.7), with the matrix of T expressed in terms of α and that of S expressed in terms of β. Find the matrices for $T \circ S$ and $S \circ T$. Does the order of composition matter?

Problem 1.4.9.

Same as (1.4.8) for T and S reflections.

Problem 1.4.10.

Repeat the analysis of (1.4.7) under the assumption that $T(1,0) = (a,b)$ with $a < 0$.

In (1.4.10) you should get two new matrices for T, again one of determinant $+1$ and one of determinant -1. Call these a **rotation of the second kind** and a **reflection of the second kind** respectively, to distinguish them from the two cases of (1.4.6). Let $\mathcal{L}^{++}(2,\mathbb{R})$ be the subset of $\mathcal{L}(2,\mathbb{R})$ consisting of rotations of the first kind.

Problem 1.4.11.

Show that the composition of two rotations of the second kind is a rotation of the first kind. Does the order of composition matter?

A group is said to be **abelian** if its group operation is commutative. (The term is in honor of Niels Abel, another famous Norwegian mathematician.) For example, $SO(2, \mathbb{R})$ is abelian, although $O(2, \mathbb{R})$ is not.

Problem 1.4.12.

a. Show that $\mathcal{L}^{++}(2, \mathbb{R})$ is a subgroup of $\mathcal{L}(2, \mathbb{R})$. (Some people call $\mathcal{L}^{++}(2, \mathbb{R})$ the Lorentz group, rather than $\mathcal{L}(2, \mathbb{R})$.)

b. Is $\mathcal{L}^{++}(2, \mathbb{R})$ abelian?

c. Do the rotations of the second kind form a subgroup of $\mathcal{L}(2, \mathbb{R})$? Why or why not?

Problem 1.4.13.

This problem relates the form of our matrices in $\mathcal{L}^{++}(2, \mathbb{R})$ to the form of those given in Taylor and Wheeler on page 42 (page 102 in the second edition), or see page 38 in Callahan.

a. Show that

$$\cosh(\alpha) = \frac{1}{\sqrt{1 - \tanh^2(\alpha)}}.$$

b. Use (a) and $\sinh(\alpha) = \cosh(\alpha)\tanh(\alpha)$ to rewrite the matrix of a rotation of the first kind using $\beta = \tanh(\alpha)$.

Problem 1.4.14.

a. Show that

$$M_\alpha = \begin{bmatrix} \cosh\alpha & \sinh\alpha & 0 & 0 \\ \sinh\alpha & \cosh\alpha & 0 & 0 \\ 0 & 0 & 1 & 0 \\ 0 & 0 & 0 & 1 \end{bmatrix}$$

is in $\mathcal{L}(4, \mathbb{R})$. (In fact, it's in $\mathcal{L}^{+}(4, \mathbb{R})$.)

b. Make up some more matrices in $\mathcal{L}(4, \mathbb{R})$, including matrices with determinant -1. Verify that your matrices really are in $\mathcal{L}(4, \mathbb{R})$.

Problem 1.4.15.

Define a 4 by 4 matrix M_A by

$$M_A = \begin{bmatrix} 1 & \mathbf{0}^T \\ \mathbf{0} & A \end{bmatrix}$$

with the column vector $\mathbf{0} \in \mathbb{R}^3$ and the matrix $A \in \mathcal{M}(3, \mathbb{R})$. Verify that M_A is in $L^{+}(4, \mathbb{R})$ if and only if A is in $SO(3, \mathbb{R})$. (You can assume that $\det(M_A) = \det(A)$.)

Remark The transformation S_α of spacetime corresponding to the matrix M_α in (1.4.14a) is called a *Lorentz boost* along the x-axis. When $A \in SO(3, \mathbb{R})$, the transformation T_A of spacetime corresponding to the matrix M_A in (1.4.15) is called a *spatial rotation*. It is a theorem that every transformation T in $\mathcal{L}^{++}(\mathbb{R}^4)$ can be written as a composition $T = T_A S_\alpha T_B$ of spatial rotations T_A and T_B and a Lorentz boost S_α. In fact, the parameter α in the Lorentz boost is uniquely determined by T. As noted in the chapter on The Geometry of Spacetime in the second (unpublished) volume of *Continuous Symmetry: From Euclid to Einstein* by Howe and Barker, "From [this] decomposition we see quite clearly that the 'relativistic information' ... is carried completely by the Lorentz boost." We'll see in Chapter 2 that transformations in $SO(\mathbb{R}^3)$ have an analogous decomposition into rotations about the coordinate axes.

Problem 1.4.16. (C)
Suppose C is an n by n matrix. Define a "dot product" on $\mathbb{R}^n$ by

$$\boldsymbol{v} \cdot \boldsymbol{w} = \boldsymbol{v}^T C \boldsymbol{w}.$$

a. What conditions on C guarantee that this dot product is a symmetric, non-degenerate bilinear form on $\mathbb{R}^n$? Verify that this is so.

b. Are your conditions necessary as well as sufficient? Justify your answer.

c. Suppose $T\colon \mathbb{R}^n \to \mathbb{R}^n$ is a linear transformation with matrix M. What condition on M is equivalent to $T(\boldsymbol{v}) \cdot T(\boldsymbol{w}) = \boldsymbol{v} \cdot \boldsymbol{w}$ for all $\boldsymbol{v}, \boldsymbol{w} \in \mathbb{R}^n$ in this case?

Another case of interest is of a non-degenerate **anti-symmetric** bilinear form on $\mathbb{R}^n$: $\boldsymbol{v} \cdot \boldsymbol{w} = -\boldsymbol{w} \cdot \boldsymbol{v}$ for all $\boldsymbol{v}, \boldsymbol{w} \in \mathbb{R}^n$.

Problem 1.4.17. (C)
Suppose C is an n by n matrix. Define a "dot product" on $\mathbb{R}^n$ by

$$\boldsymbol{v} \cdot \boldsymbol{w} = \boldsymbol{v}^T C \boldsymbol{w}.$$

a. What conditions on C guarantee that this dot product is an anti-symmetric, non-degenerate bilinear form on $\mathbb{R}^n$? Verify that this is so.

b. Are your conditions necessary as well as sufficient? Justify your answer.

c. Show that the conditions in (a) imply that n is even.

1.5 Putting the pieces together

In this section we summarize and collect in one place for reference the definitions and the theorems embedded in the problems in the preceding sections. We also mention some theorems that generalize or expand upon our results. At the end, we give some suggestions for further reading.

Definitions A nonempty set G together with an operation $*$ is a **group** provided

- The set is **closed** under the operation. That is, if g and h belong to the set G, then so does $g * h$.

- The operation is **associative**. That is, if g, h and k are any elements of G, then $g * (h * k) = (g * h) * k$.
- There is an element e of G which is an **identity** for the operation. That is, if g is any element of G, then $g * e = e * g = g$.
- Every element of G has an **inverse** in G. That is, if g is in G then there is an element of G denoted g^{-1} satisfying $g * g^{-1} = g^{-1} * g = e$.

The group G is **abelian** if $g * h = h * g$ for all $g, h \in G$.

A nonempty subset $H \subseteq G$ is a **subgroup** of the group G if H is itself a group with the same operation $*$ as G. That is, H is a subgroup provided

- H is closed under the operation $*$.
- H contains the identity e (which guarantees that H is nonempty).
- Every element of H has an inverse in H.

Many of our groups are given in two forms, one in which the group elements are linear transformations $\mathbb{R}^n \to \mathbb{R}^n$ and the group operation is composition of functions, and one in which the group elements are n by n matrices and the group operation is matrix multiplication. The "dictionary" between the language of linear transformations and that of matrices is the correspondence between a linear transformation $\mathbb{R}^n \to \mathbb{R}^n$ and its matrix with respect to the standard basis of $\mathbb{R}^n$. As we will elaborate in Chapter 3, corresponding groups in the two languages are essentially "the same" (formally, they are *isomorphic*).

1.5.1 Dot products

The familiar Euclidean dot product on $\mathbb{R}^n$ describes the *distance* between two points. For $\boldsymbol{v} = (v_1, \ldots, v_n)$, $\boldsymbol{w} = (w_1, \ldots, w_n)$, $\boldsymbol{v} \cdot \boldsymbol{w} = \sum_j v_j w_j$, and the square of the distance between points in $\mathbb{R}^n$ corresponding to $\boldsymbol{v}$ and $\boldsymbol{w}$ is $(\boldsymbol{w} - \boldsymbol{v}) \cdot (\boldsymbol{w} - \boldsymbol{v})$.

If we regard $\mathbb{R}^4$ as *spacetime*, where each vector gives the spatial and time coordinates of an event, the Lorentz dot product on $\mathbb{R}^4$ describes the *interval* between two events in special relativity. For $\boldsymbol{v}_j = (t_j, x_j, y_j, z_j) \in \mathbb{R}^4$ $(j = 1, 2)$, $\boldsymbol{v}_1 \cdot \boldsymbol{v}_2 = t_1 t_2 - x_1 x_2 - y_1 y_2 - z_1 z_2$, and the square of the interval between events $\boldsymbol{v}_1$ and $\boldsymbol{v}_2$ is $(\boldsymbol{v}_2 - \boldsymbol{v}_1) \cdot (\boldsymbol{v}_2 - \boldsymbol{v}_1)$.

More generally, we have the following definitions. A dot product (or **form**) on $\mathbb{R}^n$ is a rule that assigns a real number denoted $\boldsymbol{v} \cdot \boldsymbol{w}$ to every ordered pair of vectors $\boldsymbol{v}, \boldsymbol{w} \in \mathbb{R}^n$. The form is **bilinear** if it satisfies

$$\begin{aligned} (\boldsymbol{u} + \boldsymbol{v}) \cdot \boldsymbol{w} &= \boldsymbol{u} \cdot \boldsymbol{w} + \boldsymbol{v} \cdot \boldsymbol{w}, \\ \boldsymbol{u} \cdot (\boldsymbol{v} + \boldsymbol{w}) &= \boldsymbol{u} \cdot \boldsymbol{v} + \boldsymbol{u} \cdot \boldsymbol{w}, \\ (a\boldsymbol{v}) \cdot \boldsymbol{w} &= a(\boldsymbol{v} \cdot \boldsymbol{w}) = \boldsymbol{v} \cdot (a\boldsymbol{w}), \end{aligned}$$

for all $\boldsymbol{u}, \boldsymbol{v}, \boldsymbol{w}$ in $\mathbb{R}^n$, and all a in $\mathbb{R}$. The form is **non-degenerate** if it has the following property: If $\boldsymbol{v} \in \mathbb{R}^n$ is a particular vector satisfying

$$\boldsymbol{v} \cdot \boldsymbol{w} = 0 \quad \text{for } \underline{\text{every}} \quad \boldsymbol{w} \quad \text{in} \quad \mathbb{R}^n$$

then $\boldsymbol{v}$ is the zero vector $\boldsymbol{v} = (0, 0, \ldots, 0)$. The form is **symmetric** if $\boldsymbol{v} \cdot \boldsymbol{w} = \boldsymbol{w} \cdot \boldsymbol{v}$ for all $\boldsymbol{v}, \boldsymbol{w}$ in $\mathbb{R}^n$. The form is **anti-symmetric** if $\boldsymbol{v} \cdot \boldsymbol{w} = -\boldsymbol{w} \cdot \boldsymbol{v}$ for all $\boldsymbol{v}, \boldsymbol{w} \in \mathbb{R}^n$. The

Euclidean dot product on $\mathbb{R}^n$ is a non-degenerate, symmetric, bilinear form. The Lorentz product (on $\mathbb{R}^2$ or on $\mathbb{R}^4$) is also a non-degenerate, symmetric, bilinear form. The vector space $\mathbb{R}^n$ together with such a dot product is called a **metric vector space**. We say that a linear transformation $T\colon \mathbb{R}^n \to \mathbb{R}^n$ **preserves** this dot product if $T(\boldsymbol{v}) \cdot T(\boldsymbol{w}) = \boldsymbol{v} \cdot \boldsymbol{w}$ for all $\boldsymbol{v}, \boldsymbol{w} \in \mathbb{R}^n$, and such a transformation is called an **isometry** of $\mathbb{R}^n$. It is a symmetry of the corresponding metric vector space.

1.5.2 Groups of symmetries

Symmetries are invertible functions from some set to itself preserving some feature of the set (shape, distance, interval, ...). A set of symmetries of a set X can form a group using the operation *composition* of functions. If f and g are functions from a set X to itself, then the composition of f and g is denoted $f \circ g$, and it is defined by $(f \circ g)(x) = f(g(x))$ for x in X. The identity element for composition of functions is the function I that "doesn't do anything": $I(x) = x$ for every x in X. Composition of functions is always associative, so to determine whether a set of functions forms a group under composition it is only necessary to verify that the set is closed under composition of functions, that the set contains the identity function I, that each element of the set is invertible, and that the set contains the inverse of each of its elements.

1. The set of all symmetries preserving a particular geometric shape (e.g., a polygon, polyhedron, etc.) is a group. Often, this is a *finite* (and discrete) group.
2. The group of symmetries of the vector space $\mathbb{R}^n$ consists of all invertible linear transformations from $\mathbb{R}^n$ to itself, since linear transformations preserve the vector space operations of vector addition and scalar multiplication. The group is denoted $GL(\mathbb{R}^n)$, the **general linear group** in dimension n.
3. The group of symmetries of the Euclidean plane $\mathbb{R}^2$ —that is, the linear transformations of $\mathbb{R}^2$ that preserve Euclidean *distance*—is denoted by $O(\mathbb{R}^2)$, the **orthogonal group** in the **Euclidean** case. Equivalently, the symmetries are the linear transformations preserving the Euclidean dot product on $\mathbb{R}^2$. The group $O(\mathbb{R}^2)$ consists of rotations about the origin and reflections across lines through the origin. (See the parable of the surveyors.)
4. The group of symmetries of the Lorentz plane $\mathbb{R}^2$ —that is, the linear transformations of $\mathbb{R}^2$ which preserve the *Lorentz product* and hence the *interval*—is denoted by $\mathcal{L}(\mathbb{R}^2)$, the **Lorentz group**. (See the story of the observers in the stationary laboratory and the rocket laboratory.) The more general case, the group of symmetries of 4-dimensional spacetime, is denoted by $\mathcal{L}(\mathbb{R}^4)$.
5. The group of symmetries of $\mathbb{R}^n$ preserving a symmetric, non-degenerate bilinear form (or "dot product") on $\mathbb{R}^n$ is denoted by $O(\mathbb{R}^n)$, the general **orthogonal group**. The subgroup of transformations of determinant 1 is denoted by $SO(\mathbb{R}^n)$, the **special orthogonal group**. (This generalizes examples 3 and 4.) When the dot product is the usual Euclidean one, we say it is the **Euclidean** case. (Note that this notation doesn't specify the underlying bilinear form.)
6. The group of symmetries preserving a non-degenerate, anti-symmetric bilinear form on $\mathbb{R}^n$ (n even) is denoted $Sp\,(\mathbb{R}^n)$, the **symplectic group**. (Some authors define

the symplectic group somewhat differently, working over the complex numbers or even the quaternions.) Although the symplectic group plays a very small role in this book, it deserves mention as one of the *classical* Lie groups, and we'll have more to say about the classical Lie groups in Chapters 5 and 6.

1.5.3 Matrix groups

The elements of a matrix group are invertible n by n matrices for some positive integer n, and the group operation is matrix multiplication. The identity element is the n by n identity matrix I_n. Matrix multiplication is always associative, so to determine whether a set of n by n matrices forms a group under matrix multiplication, it is only necessary to verify that the set is closed under matrix multiplication, that the set contains I_n, that matrices in the set are invertible, and the set contains the inverse of each of its elements. These groups provide alternative descriptions of the groups listed in 1.5.2. We denote the set of all n by n matrices with real number entries by $\mathcal{M}(n, \mathbb{R})$.

1. The **general linear group** $GL(n, \mathbb{R})$ is the group of invertible n by n matrices. In other words, an n by n matrix M is in $GL(n, \mathbb{R})$ if and only if $\det(M) \neq 0$. More compactly, we can write

$$GL(n, \mathbb{R}) = \{M \in \mathcal{M}(n, \mathbb{R}) \; : \; \det(M) \neq 0\}.$$

2. The **special linear group** $SL(n, \mathbb{R})$ is the subgroup of $GL(n, \mathbb{R})$ consisting of matrices of determinant 1.

$$SL(n, \mathbb{R}) = \{M \in \mathcal{M}(n, \mathbb{R}) \; : \; \det(M) = 1\}.$$

3. The **orthogonal group** $O(2, \mathbb{R})$ in the **Euclidean** case consists of 2 by 2 matrices M satisfying $M^T M = I_2$. These matrices have the form

$$\begin{bmatrix} \cos(\alpha) & -\sin(\alpha) \\ \sin(\alpha) & \cos(\alpha) \end{bmatrix} \quad \text{or} \quad \begin{bmatrix} \cos(\alpha) & \sin(\alpha) \\ \sin(\alpha) & -\cos(\alpha) \end{bmatrix}$$

for some real number α. The linear transformation corresponding to the first matrix is a **rotation** through an angle α (counterclockwise if $\alpha > 0$ and clockwise if $\alpha < 0$). The linear transformation corresponding to the second matrix is a **reflection** across a line passing through the origin and making an angle $\alpha/2$ with the positive x axis.

4. The two-dimensional **Lorentz group** $\mathcal{L}(2, \mathbb{R})$ consists of 2 by 2 matrices M satisfying

$$M^T C M = C,$$

where the matrix C describes the Lorentz dot product with respect to the standard basis:

$$C = \begin{bmatrix} 1 & 0 \\ 0 & -1 \end{bmatrix}.$$

We can write

$$\mathcal{L}(2, \mathbb{R}) = \{M \in \mathcal{M}(2, \mathbb{R}) \; : \; M^T C M = C\}.$$

The matrices in $\mathcal{L}(2, \mathbb{R})$ are of four types, each determined by a real number α. The first two:

$$\begin{bmatrix} \cosh(\alpha) & \sinh(\alpha) \\ \sinh(\alpha) & \cosh(\alpha) \end{bmatrix} \quad \text{or} \quad \begin{bmatrix} \cosh(\alpha) & -\sinh(\alpha) \\ \sinh(\alpha) & -\cosh(\alpha) \end{bmatrix}$$

are, respectively, **rotations and reflections of the first kind**.

The second two:

$$\begin{bmatrix} -\cosh(\alpha) & \sinh(\alpha) \\ \sinh(\alpha) & -\cosh(\alpha) \end{bmatrix} \quad \text{or} \quad \begin{bmatrix} -\cosh(\alpha) & -\sinh(\alpha) \\ \sinh(\alpha) & \cosh(\alpha) \end{bmatrix}$$

are, respectively, **rotations and reflections of the second kind**. $\mathcal{L}(2, \mathbb{R})$ has two subgroups of interest, $\mathcal{L}^{+}(2, \mathbb{R})$ is the group of rotations (of both kinds), and $\mathcal{L}^{++}(2, \mathbb{R})$ is the group of rotations of the first kind.

5. The **orthogonal group** $O(n, \mathbb{R})$ consists of n by n matrices M satisfying

$$M^T C M = C,$$

where the matrix C is symmetric and invertible. We can write

$$O(n, \mathbb{R}) = \{M \in \mathcal{M}(n, \mathbb{R}) \,:\, M^T C M = C\}.$$

In the Euclidean case, $C = I_n$, the identity matrix. In the case of $\mathcal{L}(2, \mathbb{R})$,

$$C = \begin{bmatrix} 1 & 0 \\ 0 & -1 \end{bmatrix}.$$

In the case of $\mathcal{L}(4, \mathbb{R})$,

$$C = \begin{bmatrix} 1 & 0 & 0 & 0 \\ 0 & -1 & 0 & 0 \\ 0 & 0 & -1 & 0 \\ 0 & 0 & 0 & -1 \end{bmatrix}.$$

6. The **symplectic group** $Sp(n, \mathbb{R})$ (for n even) consists of n by n matrices M satisfying $M^T C M = C$, where the matrix C is anti-symmetric ($C^T = -C$) and invertible.

7. The **rotation subgroup** or **special orthogonal group** $SO(n, \mathbb{R})$ consists of those elements of the orthogonal group $O(n, \mathbb{R})$ which have determinant 1. The rotation subgroup is also sometimes denoted $O^{+}(n, \mathbb{R})$.

$$SO(n, \mathbb{R}) = \{M \in O(n, \mathbb{R}) \,:\, \det(M) = 1\}.$$

Also see the definitions of the **unitary group** and the **special unitary group** in the next chapter.

Remarks

- By *Sylvester's Theorem*, every finite-dimensional real vector space with a non–degenerate symmetric bilinear form has a basis with respect to which the matrix C describing the form is diagonal, with diagonal entries ± 1. The number r of $+1$s is called the *signature* of the form. The corresponding orthogonal group is often denoted by $O(r, n-r, \mathbb{R})$. So, for example, the Lorentz group $\mathcal{L}(4, \mathbb{R})$ would be denoted by $O(1, 3, \mathbb{R})$. (Note that in this example, replacing C by $-C$ gives a form with a different signature, even though the group of isometries is unchanged.)
- Every finite-dimensional real vector space with a non-degenerate anti-symmetric bilinear form has even dimension $n = 2m$. By a result analogous to Sylvester's Theorem, such a space has a basis with respect to which the matrix C describing the form is

$$C = \begin{bmatrix} 0_m & I_m \\ -I_m & 0_m \end{bmatrix},$$

where 0_m is the m by m all-zero matrix.

1.5.4 Theorems

The proofs of the following theorems are embedded in the problems of sections 1.2–1.4.

Theorem 1.1

a. ***Transformation version.*** *For any positive integer n, $GL(\mathbb{R}^n)$ is a group under composition of functions, and $SL(\mathbb{R}^n)$ is a subgroup of $GL(\mathbb{R}^n)$.*

b. ***Matrix version.*** *For any positive integer n, $GL(n, \mathbb{R})$ is a group under matrix multiplication, and $SL(n, \mathbb{R})$ is a subgroup of $GL(n, \mathbb{R})$.*

Theorem 1.2 *Fix a positive integer n. Assume $T\colon \mathbb{R}^n \to \mathbb{R}^n$ is a linear transformation and that M is the matrix of T with respect to the standard basis of $\mathbb{R}^n$. Then the following conditions (a)–(c) are equivalent.*

a. *T preserves distance; that is, for any points P and Q in $\mathbb{R}^n$, the distance between P and Q is the same as the distance between $T(P)$ and $T(Q)$.*

b. *T preserves the (Euclidean) dot product; that is, for any vectors $\boldsymbol{v}$, $\boldsymbol{w}$ in $\mathbb{R}^n$, $\boldsymbol{v} \cdot \boldsymbol{w} = T(\boldsymbol{v}) \cdot T(\boldsymbol{w})$.*

c. $M^T M = I_n$.

d. *For $n = 2$, if T satisfies (a)–(c), then T is either a rotation or a reflection, and M has one of the two forms in example 3 of subsection 1.5.3.*

Theorem 1.3 *Assume $T\colon \mathbb{R}^2 \to \mathbb{R}^2$ is a linear transformation and that M is the matrix of T with respect to the standard basis of $\mathbb{R}^2$. Then the following conditions are equivalent.*

a. *T preserves the interval; that is, for any events P and Q in $\mathbb{R}^2$, the interval between P and Q is the same as the interval between $T(P)$ and $T(Q)$.*

b. *T preserves the Lorentz dot product; that is, for any vectors $\boldsymbol{v}$, $\boldsymbol{w}$ in $\mathbb{R}^2$, $\boldsymbol{v} \cdot \boldsymbol{w} = T(\boldsymbol{v}) \cdot T(\boldsymbol{w})$.*

c. *$M^T CM = C$ for the 2 by 2 matrix C in example 4 of subsection 1.5.3.*

d. *T is either a rotation or a reflection and of the first or second kind, and M has one of the four forms in example 4 of subsection 1.5.3.*

Theorem 1.4 *Fix a positive integer n. Assume $T: \mathbb{R}^n \to \mathbb{R}^n$ is a linear transformation, and assume M is the matrix of T with respect to the standard basis. Assume also that $\boldsymbol{v} \cdot \boldsymbol{w}$ denotes a non-degenerate, symmetric bilinear form on $\mathbb{R}^n$ and that C is the n by n matrix for which $\boldsymbol{v} \cdot \boldsymbol{w} = \boldsymbol{v}^T C \boldsymbol{w}$. Then the following conditions are equivalent.*

a. *T preserves the bilinear form; that is, for any vectors $\boldsymbol{v}$, $\boldsymbol{w}$ in $\mathbb{R}^n$,*

$$\boldsymbol{v} \cdot \boldsymbol{w} = T(\boldsymbol{v}) \cdot T(\boldsymbol{w}).$$

b. $M^T CM = C$.

Theorem 1.5

a. ***Transformation version.*** *For any positive integer n, $O(\mathbb{R}^n)$ is a group under composition of functions, and $SO(\mathbb{R}^n)$ is a subgroup of $O(\mathbb{R}^n)$.*

b. ***Matrix version.*** *For any positive integer n, $O(n, \mathbb{R})$ is a group under matrix multiplication, and $SO(n, \mathbb{R})$ is a subgroup of $O(n, \mathbb{R})$.*

Suggestions for further reading

Singer, S.F., *Symmetry in Mechanics: A Gentle, Modern Introduction*, Birk-häuser (2001). For students with background in physics, section 2.2 gives a context for the use of an anti-symmetric bilinear form.

Snapper, E. and Troyer, R.J., *Metric Affine Geometry*, Academic Press (1971). Chapter 2 on Metric Vector Spaces takes a more detailed look at the geometry and isometries of a vector space with a symmetric bilinear form, including Sylvester's Theorem. It also contains some interesting asides for prospective high school teachers.

1.6 A broader view: Lie groups

So, what *is* a "Lie group"? The formal definition of a Lie group uses ideas from analysis and topology, beyond the prerequisites for this book. Very roughly speaking, a Lie group is a group on which we can do calculus. (More formally, a Lie group has the structure of a "differentiable manifold.") Further, the group operation $(M_1, M_2) \mapsto M_1 * M_2$ and inversion $M \mapsto M^{-1}$ are required to be differentiable functions from this calculus point of view. (In fact, they are required to be *smooth*, meaning that they are "infinitely" differentiable: second, third, ..., nth, ... derivatives all exist.)

We will mostly restrict ourselves to groups of matrices — initially, subgroups of $GL(n, \mathbb{R})$. Many Lie groups are matrix groups, but not all. (A major theorem, called the Peter-Weyl theorem, has as a corollary that every "compact" Lie group is a matrix group.) Most of the Lie groups with important applications in science are matrix groups.

However, not every matrix group is a Lie group. Only certain subgroups of $GL(n, \mathbb{R})$ qualify as Lie groups, namely the ones that are "closed sets" in the topology on the group. A closed set is analogous to a closed interval on the real line. More formally, a set is closed if it contains the limit of any convergent sequence formed of elements of the set. (In Chapter 4 we will look at what it means for a sequence of matrices to converge.) The classical matrix groups $GL(n, \mathbb{R})$, $SL(n, \mathbb{R})$, $O(n, \mathbb{R})$ (all cases), and $SO(n, \mathbb{R})$ are all Lie groups, as is $Sp(n, \mathbb{R})$ for even n. Also, $\mathbb{R}$ and (for a fixed positive integer n) the vector space $\mathbb{R}^n$ are groups under addition, and the nonzero elements of $\mathbb{R}$ form a group under multiplication; all of these are Lie groups. (In the next chapter we'll add more examples to our list: the complex numbers, matrices with complex entries, and a number system called the "quaternions.")

We can identify $\mathbb{Q}$ with a group of matrices G, where

$$G = \left\{ \begin{bmatrix} 1 & a \\ 0 & 1 \end{bmatrix} : a \in \mathbb{Q} \right\}.$$

Using language we'll introduce in Chapter 3, G with operation matrix multiplication and $\mathbb{Q}$ with operation addition are "isomorphic" groups, and therefore essentially the same. But $\mathbb{Q}$ is not a Lie group because it's not a manifold, so G isn't either. Also $\mathbb{Q}$ is not a closed subgroup of $\mathbb{R}$, because there are sequences of rational numbers that converge to irrational numbers. (Can you think of an example?) So G isn't a closed subgroup of $GL(2, \mathbb{R})$.

In this book, unless specified otherwise, when we write "Lie group" we mean matrix Lie group. Even with our restriction to matrix groups, we'll be able to work with many of the key ideas in the theory and applications of Lie groups.

CHAPTER 2

Complex numbers, quaternions and geometry

This chapter introduces two number systems that will play important roles in our study of Lie groups, the *complex numbers* $\mathbb{C}$ and the *quaternions* $\mathbb{H}$. (The letter H for the quaternions is in honor of the Irish mathematician William Rowan Hamilton, who discovered them in 1862. The discovery is a wonderful story. See *On Quaternions and Octonions* by Conway and Smith, pp. 6–10; or use a search engine to look for Hamilton and quaternions to read about it.) As we shall see, the complex numbers give us another way to view the Euclidean geometry and symmetries of $\mathbb{R}^2$, and the quaternions give another view of the geometry and symmetries of $\mathbb{R}^3$. This chapter also gives more practice with the definition of a group.

2.1 Complex numbers

The **complex numbers** are

$$\mathbb{C} = \{a + bi : a, b \in \mathbb{R}\},$$

where addition and multiplication are defined by $i^2 = -1$ and the usual rules in $\mathbb{R}$. That is,

$$\begin{aligned}(a + bi) + (a' + b'i) &= (a + a') + (b + b')i,\\ (a + bi)(a' + b'i) &= (aa' - bb') + (ab' + ba')i.\end{aligned}$$

It is straightforward to verify that addition and multiplication are associative and commutative and satisfy the distributive law.

- We think of the real numbers as a subset of the complex numbers: $a + bi$ is real if and only if $b = 0$.
- A complex number $a + bi$ is zero if and only if $a = b = 0$.
- Two complex numbers $a + bi$ and $c + di$ are equal if and only if $a = c$ and $b = d$.

Another kind of operation on $\mathbb{C}$ is **complex conjugation**, where the complex conjugate of $z = a + bi$ is

$$\overline{z} = a - bi.$$

Technically, addition and multiplication are *binary* operations—you add or multiply *two* numbers, while conjugation is a *unary* operation—you find the complex conjugate of *one* number.

Problem 2.1.1.
Verify the following facts about a complex number $z = a + bi$.

a. $\overline{z} = z$ if and only if z is a real number. Also, for any z, $z\overline{z}$ is a nonnegative real number, and $z\overline{z} = 0$ if and only if $z = 0$.

b. If we interpret z as a vector (directed line segment) from the origin to (a, b), then the length of that vector is $|z| = \sqrt{z\overline{z}}$.

c. A nonzero complex number z has a multiplicative inverse z^{-1} satisfying $zz^{-1} = z^{-1}z = 1$, namely

$$z^{-1} = \frac{1}{z} = \left(\frac{1}{z\overline{z}}\right)\overline{z} = \left(\frac{a}{a^2+b^2}\right) - \left(\frac{b}{a^2+b^2}\right)i.$$

Problem 2.1.2.
In this problem, z, z_1 and z_2 are arbitrary complex numbers.

a. Verify that $\overline{\overline{z}} = z$, $\overline{z_1 + z_2} = \overline{z}_1 + \overline{z}_2$, $\overline{z_1 z_2} = \overline{z}_1\overline{z}_2$, and $\overline{1/z} = 1/\overline{z}$.

b. Verify that $|z_1 z_2| = |z_1||z_2|$ and $|\overline{z}| = |z|$.

Problem 2.1.3.

a. Show the complex numbers form a group under addition. You have to check three things:

- Show that $\mathbb{C}$ is **closed** under addition—the sum of two complex numbers is again a complex number.
- Show that there is an **identity** for addition in $\mathbb{C}$. (Explain why 0 is the right choice for the identity for addition, i.e., the analog of I for composition of transformations or I_n for multiplication of matrices.)
- Show that every complex number has an additive **inverse** which is again a complex number. (What is the *inverse* of a complex number for addition?)

b. Show the *nonzero* complex numbers form a group under multiplication. You have to check three things:

- Show that $\mathbb{C}$ is **closed** under multiplication—the product of two nonzero complex numbers is again a nonzero complex number.
- Show that there is an **identity** for multiplication among the nonzero complex numbers. (Explain why 1 is the right choice for the identity for multiplication, i.e., the analog of I for composition of transformations or I_n for multiplication of matrices.)
- Show that every nonzero complex number has a multiplicative **inverse** which is again a nonzero complex number.

A standard way to draw a picture of a complex number is to represent $a + bi$ by the point (or, equivalently, the vector) $(a, b) \in \mathbb{R}^2$. In this representation, the horizontal axis is the *real* axis, and the vertical is *imaginary*. As in (2.1.1b), the length of the vector corresponding to $z = a + bi$ is then $\sqrt{z\bar{z}}$. Notice that a complex number of "length" 1 is thus one of the form $\cos(\alpha) + i\sin(\alpha)$ for some real number α. (See Figure 2.1.)

Recall from 1.4 the Taylor series (centered at 0) for the exponential function and for the sine and cosine.

Problem 2.1.4.

a. Substitute $x = i\alpha$ into the series for e^x and, comparing to the series for sine and cosine, show that $e^{i\alpha} = \cos(\alpha) + i\sin(\alpha)$.

b. Use (a) to show $e^0 = e^{2\pi i} = 1$. What is $e^{\pi i}$?

c. Use (a) and the rules for multiplying complex numbers and trigonometric identities to verify that $e^{i\alpha}e^{i\beta} = e^{i\alpha+i\beta} = e^{i(\alpha+\beta)}$. (Don't just *assume* that the usual rules for exponents extend to complex exponents—that's what you're proving!)

d. Show that the complex conjugate of $e^{i\alpha}$ is $e^{-i\alpha}$.

e. Explain why $e^{i\alpha}$ is never equal to zero. What is the multiplicative inverse of $e^{i\alpha}$?

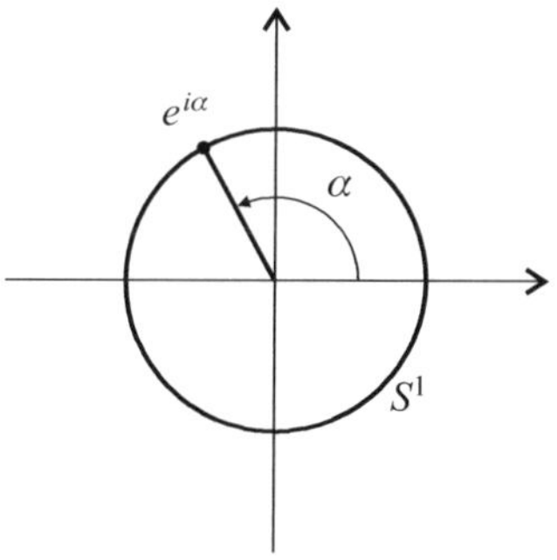

Figure 2.1. The unit circle S^1 and $e^{i\alpha}$ in S^1

In a natural way, each complex number $e^{i\alpha}$ can be identified with a point of the unit circle S^1 in $\mathbb{R}^2$. (See Figure 2.1.) The unit circle is sometimes called a **1-sphere** because it is 1-dimensional, a curve—that's the reason for the exponent 1 in S^1. Conversely, each point of the unit circle can be identified with a complex number of the form $e^{i\alpha}$:

$$S^1 = \{e^{i\alpha} : \alpha \in \mathbb{R}\} = \{e^{i\alpha} : 0 \le \alpha < 2\pi\} = \{z \in \mathbb{C} : |z| = 1\}.$$

Remark Both $\{e^{i\alpha} : \alpha \in \mathbb{R}\}$ and $\{e^{i\alpha} : 0 \le \alpha < 2\pi\}$ describe the same set of numbers. The periodicity of the sine and cosine means that each complex number (or point in $\mathbb{R}^2$) is named infinitely often in the first description, but the redundant naming does not change which numbers are in the set.

Problem 2.1.5.

Use the description you prefer (and previous results, as needed) to show that S^1 is a group under the usual multiplication of complex numbers. That is, show

- S^1 is **closed** under multiplication: if z_1 and z_2 belong to S^1, then so does their product z_1z_2.
- The complex number 1 is in S^1.
- If z is in S^1, then its multiplicative **inverse** z^{-1} is also in S^1.

Problem 2.1.6. (S)
Fix the real number α. Define a function $T_\alpha : \mathbb{C} \to \mathbb{C}$ by

$$T_\alpha(z) = e^{i\alpha} z \quad \text{for } z \in \mathbb{C}.$$

a. Show that T_α preserves "lengths" of complex numbers. That is, show $|T_\alpha(z)| = |z|$ for any $z \in \mathbb{C}$.
b. Interpreting complex numbers as vectors in $\mathbb{R}^2$, show that T_α is a linear transformation. That is, show $T_\alpha(z_1+z_2) = T_\alpha(z_1)+T_\alpha(z_2)$ and $T_\alpha(rz) = rT_\alpha(z)$ for any $z_1, z_2, z \in \mathbb{C}$ and $r \in \mathbb{R}$. (Note that (a) and (b) together tell us that T_α is an **isometry** of $\mathbb{C}$.)
c. What is the matrix of T_α with respect to the standard basis of $\mathbb{R}^2$? (Have you seen this matrix before?)

Problem 2.1.7.
Again fix the real number α. This time define a function $S_\alpha : \mathbb{C} \to \mathbb{C}$ by

$$S_\alpha(z) = e^{i\alpha}\, \overline{z} \quad \text{for } z \in \mathbb{C}.$$

a. Show that S_α preserves lengths of complex numbers.
b. Interpreting complex numbers as vectors in $\mathbb{R}^2$, show that S_α is a linear transformation. (So, by (a) and (b), S_α is another isometry of $\mathbb{C}$.)
c. What is the matrix of S_α with respect to the standard basis of $\mathbb{R}^2$? (Have you seen this matrix before?)

We'll have occasion to use matrices with complex entries: $\mathcal{M}(n,\mathbb{C})$ is the set of all n by n matrices with entries in $\mathbb{C}$. If the matrix $A \in \mathcal{M}(n,\mathbb{C})$ has entries a_{ij}, write $\overline{A}$ for the matrix with entries $\overline{a}_{ij}$. (Similarly, the complex conjugate of a vector $\boldsymbol{v} = (v_1, \ldots, v_n)$ with complex entries is $\overline{\boldsymbol{v}} = (\overline{v}_1, \ldots, \overline{v}_n)$.) Complex conjugation has a simple relationship to matrix operations on $A, B \in \mathcal{M}(n,\mathbb{C})$, as the following problem shows.

Problem 2.1.8.
For each of the following, you may make your argument for the case when $n = 2$, but to make the generalization to arbitrary n easier, write your 2 by 2 matrices in the form

$$A = \begin{bmatrix} a_{11} & a_{12} \\ a_{21} & a_{22} \end{bmatrix}.$$

a. Show $\overline{A+B} = \overline{A} + \overline{B}$.
b. Show that $\overline{AB} = \overline{A}\,\overline{B}$.
c. Show $\overline{A^T} = (\overline{A})^T$.
d. Show $\mathrm{tr}(\overline{A}) = \overline{\mathrm{tr}(A)}$.
e. Show $\det(\overline{A}) = \overline{\det(A)}$.

This is getting a bit ahead of our story, but there's a very important group of matrices with complex entries called the **unitary group** $U(n,\mathbb{C})$.

$$U(n,\mathbb{C}) = \{M \in \mathcal{M}(n,\mathbb{C}) : \overline{M}^T M = I_n\}.$$

It has an important subgroup, the **special unitary group** $SU(n,\mathbb{C})$.

$$SU(n,\mathbb{C}) = \{M \in U(n,\mathbb{C}) : \det(M) = 1\}.$$

Problem 2.1.9.

Assume the results of (2.1.8) for the following. Also assume that the usual rules for matrix multiplication, transposes and determinants hold for matrices with complex entries.

a. Show that $U(n, \mathbb{C})$ is a group under matrix multiplication.

b. Show that $SU(n, \mathbb{C})$ is a subgroup of $U(n, \mathbb{C})$.

c. Show that if $M \in SU(2, \mathbb{C})$, then

$$M = \begin{bmatrix} u & -\overline{v} \\ v & \overline{u} \end{bmatrix}$$

for complex numbers u and v satisfying $u\overline{u} + v\overline{v} = 1$. (Hint: Set up notation

$$M = \begin{bmatrix} u & w \\ v & z \end{bmatrix},$$

and write out the equations involving u, v, w, z that correspond to the requirements $\overline{M}^T M = I_2$ and $\det(M) = 1$. Use the equations to write w and z in terms of u and v, but consider two cases separately:
Case (i) $u = 0$, and
Case (ii) $u \neq 0$.)

In Chapter 6 we'll motivate the definition of the unitary group more fully and do more with it. For some motivation now, we mention the importance of $SU(2, \mathbb{C})$ in physics. For example, Singer points out on page 117 of her book *Linearity, Symmetry, and Prediction in the Hydrogen Atom* that physicists use exactly four groups to make predictions about the "electron shells" and "energy levels" of a hydrogen atom: $SO(2, \mathbb{R})$, $SO(3, \mathbb{R})$, $SO(4, \mathbb{R})$ and our new group $SU(2, \mathbb{C})$.

2.2 Quaternions

The **quaternions** $\mathbb{H}$ are an extension of the complex numbers:

$$\mathbb{H} = \{a + bi + cj + dk \, : \, a, b, c, d \in \mathbb{R}\},$$

where $i^2 = j^2 = k^2 = -1$, $ij = k = -ji$, $jk = i = -kj$, and $ki = j = -ik$. Otherwise, addition and multiplication follow the usual rules for $\mathbb{R}$: for $z = a + bi + cj + dk$ and $z' = a' + b'i + c'j + d'k$,

$$\begin{aligned} z + z' &= (a + a') + (b + b')i + (c + c')j + (d + d')k, \\ zz' &= (aa' - bb' - cc' - dd') + (ab' + ba' + cd' - dc')i \\ &\quad + (ac' + ca' + db' - bd')j + (ad' + da' + bc' - cb')k. \end{aligned}$$

Addition and multiplication are associative and satisfy the distributive law. Addition is commutative, but multiplication in general is *not*. (However, see (2.2.1).)

- $\mathbb{R} \subset \mathbb{C} \subset \mathbb{H}$; the quaternion $a + bi + cj + dk$ is real if and only if $b = c = d = 0$ and complex if and only if $c = d = 0$.
- $a + bi + cj + dk = 0$ if and only if $a = b = c = d = 0$.
- $a + bi + cj + dk = a' + b'i + c'j + d'k$ if and only if $a = a'$, $b = b'$, $c = c'$ and $d = d'$.

Problem 2.2.1.

a. Suppose $r \in \mathbb{R}$ and $z \in \mathbb{H}$. Show $rz = zr$.

b. Is the converse of (a) true? In other words, if $x \in \mathbb{H}$ and x commutes with every z in $\mathbb{H}$, does x have to be real?

An analogue of complex conjugation is also defined for $\mathbb{H}$. The **quaternionic conjugate** of $z = a + bi + cj + dk$ is $\overline{z} = a - bi - cj - dk$.

Problem 2.2.2.

a. Show $\overline{z_1 + z_2} = \overline{z}_1 + \overline{z}_2$ and $\overline{\overline{z}} = z$ for all $z, z_1, z_2 \in \mathbb{H}$.

b. **(S)** Calculate $z\overline{z}$, show $z\overline{z}$ is a nonnegative real number, and find a formula for the multiplicative inverse $1/z$ for $z \in \mathbb{H}$, $z \neq 0$.

c. Show $\mathbb{H}$ is a group under addition.

d. **(S)** Show the *nonzero* elements of $\mathbb{H}$ form a group under multiplication.

e. Since multiplication in $\mathbb{H}$ isn't commutative, perhaps it's not surprising that the relationship between multiplication and complex conjugation is different for $\mathbb{H}$ than for $\mathbb{C}$. Show $\overline{z_1 z_2} = \overline{z}_2\, \overline{z}_1$ for all $z_1, z_2 \in \mathbb{H}$. (Hint: you can somewhat simplify an otherwise messy calculation by writing $z = a + w$ for a real and $w = bi + cj + dk$ and observing $\overline{z} = a - w$.)

f. Show $\{z \in \mathbb{H} : z\overline{z} = 1\}$ is a group under multiplication (so it is a subgroup of the multiplicative group of nonzero quaternions).

g. Define $|z| = \sqrt{z\overline{z}}$. Show $|z_1 z_2| = |z_1||z_2|$ for all $z_1, z_2 \in \mathbb{H}$.

If $z \in \mathbb{H}$ is identified with the vector $(a, b, c, d) \in \mathbb{R}^4$, then $|z|$ is its Euclidean length. The group in (2.2.2f) can thus be viewed as the set of vectors in $\mathbb{R}^4$ of length 1 and is often called the **3-sphere** S^3 in $\mathbb{R}^4$, by analogy to the 1-sphere S^1 in $\mathbb{R}^2$,

$$S^3 = \{q \in \mathbb{H} : q\overline{q} = 1\}.$$

The 3-sphere, like the 1-sphere, is a group: the 1-sphere is a subgroup of the multiplicative group of nonzero complex numbers, and the 3-sphere is a subgroup of the multiplicative group of nonzero quaternions. The **2-sphere** S^2 is the surface of the "usual" unit sphere in $\mathbb{R}^3$. It turns out that S^2 does not have a natural group structure. In fact, it is a theorem that the *only* spheres—in any dimension—on which continuous group operations can be defined are S^3 and S^1, so these examples are very, very special.

One of the important applications of the quaternions is in computer graphics, for the efficient description of rotations in 3-dimensional space (see section 2.3). The results of the next two problems will be useful in that context.

Problem 2.2.3

If $z \in \mathbb{H}$ has real part zero (i.e., if z satisfies $z + \overline{z} = 0$), we can interpret z as a vector in $\mathbb{R}^3$, identifying $xi + yj + zk \in \mathbb{H}$ with $(x, y, z) \in \mathbb{R}^3$.

a. For a quaternion z with real part zero, show that $|z|$ is the length of z regarded as a vector in $\mathbb{R}^3$.

b. Suppose $z_1, z_2 \in \mathbb{H}$ each have real part zero. Show that $z_1 z_2 = -(z_1 \cdot z_2) + (z_1 \times z_2)$. (On the left side of this equation is the multiplication in $\mathbb{H}$, and on the right are the dot product and cross product operations on $\mathbb{R}^3$.)

Problem 2.2.4

a. Let $q \in \mathbb{H}$ have the form $q = \cos(\alpha/2) + \sin(\alpha/2)u$, where $\alpha \in \mathbb{R}$ and $u \in S^3$ has real part zero. Calculate $\overline{q}$, $q\overline{q}$ and q^{-1}.

b. **(C)** Show that every $q \in S^3$ has the form in (a) for some choice of $\alpha \in \mathbb{R}$ and some $u \in S^3$ with real part zero.

2.3 The geometry of rotations of $\mathbb{R}^3$

In this section we continue with the Euclidean dot product. Our goal is a geometric description of the transformations in $SO(\mathbb{R}^3)$. We will show that these rotations deserve their name. But first, we need to specify what we mean geometrically by a rotation of $\mathbb{R}^3$. A rotation of the plane spins around a point; a **rotation of three-space** spins around a line called the **axis of rotation**—think of a globe representing the earth and spinning around the axis through the north and south poles, for example. We'll only consider rotations of $\mathbb{R}^3$ whose axes are lines through the origin.

Visualizing transformations in three dimensions isn't easy. An old tennis ball can be a useful aid to thinking about rotations of $\mathbb{R}^3$. Imagine the tennis ball is centered at the origin, and assume it's a perfect sphere of radius 1. Now imagine an axis through the origin. With a pen, you can mark the points on the ball where the axis would pierce its surface. Rotating all of $\mathbb{R}^3$ around the axis has the effect of rotating the ball around the axis. (Do you see why a point on the surface of the ball will rotate to another point on the surface of the ball? If not, look back at problem (1.2.1).) You can mark a point of your choice on the tennis ball and then see where it moves when you rotate the ball through some angle around the imagined axis. This visual aid can be especially helpful in thinking about problems (2.3.6)–(2.3.8) in this section.

Problem 2.3.1. (S)

Answer these questions by reasoning geometrically.

a. Start in two dimensions and suppose T is a rotation of $\mathbb{R}^2$ about the origin. Recall that the nonzero vector $\boldsymbol{v}$ is an *eigenvector* for the linear transformation T with *eigenvalue* $\lambda \in \mathbb{R}$ if $T(\boldsymbol{v}) = \lambda\boldsymbol{v}$. What are the eigenvector(s) and eigenvalue(s) of T (if any)? Does the size of the angle of the rotation matter?

b. Still in two dimensions, suppose T is a reflection of $\mathbb{R}^2$ across a line through the origin spanned by the nonzero vector $\boldsymbol{a}$. What are the eigenvector(s) and eigenvalue(s) of T (if any)? Does the choice of the mirror line matter?

c. Now, return to three dimensions and suppose T is a rotation of $\mathbb{R}^3$ around an axis spanned by the nonzero vector $\boldsymbol{a}$. What are the eigenvector(s) and eigenvalue(s) of T (if any)? Does the size of the angle of the rotation matter? Does the choice of the axis matter?

Problem 2.3.2.

Assume $T \in SO(\mathbb{R}^3)$ and T has matrix $M \in SO(3, \mathbb{R})$ with respect to the usual basis of $\mathbb{R}^3$.

a. Justify each line of the following argument.

$$\begin{aligned}\det(M - I_3) &= \det(M - M^T M) \\ &= \det((I_3 - M^T)M) \\ &= \det(I_3 - M^T)\det(M) \\ &= \det(I_3 - M^T) \\ &= \det((I_3 - M)^T) \\ &= \det(I_3 - M) \\ &= (-1)^3 \det(M - I_3).\end{aligned}$$

b. Explain why (a) tells you that $\det(M - I_3) = 0$.

c. Explain why (b) tells you that T has an eigenvector with eigenvalue $+1$.

Problem 2.3.3.

Suppose T is in $O(\mathbb{R}^3)$ and $\det(T) = -1$. Use a variant of the argument in (2.3.2) to explain why T has an eigenvector with eigenvalue -1.

Problem 2.3.4.

Let $T \in SO(\mathbb{R}^3)$.

a. Choose a fixed nonzero vector $\boldsymbol{a} \in \mathbb{R}^3$. Give a geometric description of

$$U = \{\boldsymbol{u} \in \mathbb{R}^3 : \boldsymbol{u} \cdot \boldsymbol{a} = 0\}.$$

b. Keeping the notation of (a), suppose $\boldsymbol{a}$ is an eigenvector for T with eigenvalue $+1$. Explain why $T(U) = \{T(\boldsymbol{u}) : \boldsymbol{u} \in U\}$ must be exactly the same subspace as U.

c. Explain why T preserves distances between points of U.

d. **(S)** Consider the special case $\boldsymbol{a} = (0, 0, 1)$. What is U in this case? Think of $T: U \to U$ as if it were $T: \mathbb{R}^2 \to \mathbb{R}^2$, a linear transformation preserving distances. Adapting our results from section 1.2, we know that the matrix of T with respect to the standard basis of $\mathbb{R}^3$ has one of two forms, depending on how T affects the vectors of U. Write these matrices. What is the determinant of each? Which can be the matrix of T?

e. Continuing (d), describe the geometric effect of T on U, and then describe the geometric effect of T on $\mathbb{R}^3$.

Problem 2.3.5.

Use (2.3.4) to help you explain why every element of $SO(\mathbb{R}^3)$ has a geometric description as a rotation about some axis in $\mathbb{R}^3$.

Try using a tennis ball or a comparable visual aid for the next three problems.

Problem 2.3.6.

Suppose linear transformations T_1 and T_2 have the matrices M_1 and M_2 with respect to the standard bases of $\mathbb{R}^3$, with

$$M_1 = \begin{bmatrix} \cos(\alpha_1) & -\sin(\alpha_1) & 0 \\ \sin(\alpha_1) & \cos(\alpha_1) & 0 \\ 0 & 0 & 1 \end{bmatrix}, \quad M_2 = \begin{bmatrix} \cos(\alpha_2) & 0 & -\sin(\alpha_2) \\ 0 & 1 & 0 \\ \sin(\alpha_2) & 0 & \cos(\alpha_2) \end{bmatrix}.$$

a. (S) Show that M_1 and M_2 are in $SO(3, \mathbb{R})$.
b. (S) Describe T_1 and T_2 geometrically.
c. (S) Suppose (a, b, c) is a point at distance 1 from the origin in $\mathbb{R}^3$. Describe geometrically how you would choose the angles α_1 and α_2 so that $T_2(T_1(a, b, c)) = (0, 0, 1)$.
d. (C) Continuing (c), assume (a, b, c) is in the first octant and write expressions for $\sin(\alpha_j)$ and $\cos(\alpha_j)$ $(j = 1, 2)$ in terms of a, b and c.

Problem 2.3.7.
Suppose $T \in SO(\mathbb{R}^3)$ has eigenvector $\boldsymbol{a}$ with eigenvalue $+1$. Choose $\boldsymbol{a}$ with $|\boldsymbol{a}| = 1$.
a. How does the result of (2.3.6) tell you that you can choose a transformation $S = T_2T_1 \in SO(\mathbb{R}^3)$ with $S(\boldsymbol{a}) = (0, 0, 1)$?
b. Is STS^{-1} in $SO(\mathbb{R}^3)$? Explain.
c. Calculate $STS^{-1}(0, 0, 1)$.
d. Explain why STS^{-1} is a rotation around the z-axis, and write a matrix for STS^{-1} with respect to the standard basis. (An unspecified angle β will appear in your matrix.)

Problem 2.3.8.
Continue as in (2.3.7), but with $\boldsymbol{a} = (\sqrt{3}/3, \sqrt{3}/3, \sqrt{3}/3)$, as follows. All matrices are with respect to the standard basis.
a. (S) Find the matrix of a transformation $S \in SO(\mathbb{R}^3)$ with $S(\boldsymbol{a}) = (0, 0, 1)$.
b. Find a matrix for STS^{-1}.
c. Use the matrices for STS^{-1} and for S to calculate a matrix for T in terms of β. (Hint: $T = S^{-1}(STS^{-1})S$.)

Problem 2.3.9.
Repeat (2.3.8) for $\boldsymbol{a} = (\sqrt{1/6}, \sqrt{1/3}, \sqrt{1/2})$.

Problem 2.3.10.
a. Find a *diagonal* matrix D_0 in $O(3, \mathbb{R})$ but *not* in $SO(3, \mathbb{R})$.
b. Show that $O(3, \mathbb{R}) = SO(3, \mathbb{R}) \cup SO(3, \mathbb{R})D_0$, where

$$SO(3, \mathbb{R})D_0 = \{M'D_0 : M' \in SO(3, \mathbb{R})\}.$$

In other words, if $M \in O(3, \mathbb{R})$ but $M \notin SO(3, \mathbb{R})$, find a matrix $M' \in SO(3, \mathbb{R})$ with $M = M'D_0$. (The description of M' will, of course, depend on M.)
c. Let $T_0 \in O(\mathbb{R}^3)$ be the linear transformation with matrix D_0 with respect to the standard basis of $\mathbb{R}^3$. Give the transformation version of your argument in (b). That is, show that $O(\mathbb{R}^3) = SO(\mathbb{R}^3) \cup SO(\mathbb{R}^3)T_0$.
d. How is the result in (c) related to the results of (2.1.6) and (2.1.7)?

We know that the composition of two rotations is again a rotation. It should be possible to determine the axis of rotation for the composition of two rotations of $\mathbb{R}^3$. This is easy if the two rotations have the *same* axis. (Why is it easy?) It can be complicated if the two rotations have different axes. Unless you're *very* unusual, using a visual aid like a tennis ball won't give the entire answer to the following problem, but it could help you figure out the matrices of S and T with respect to the standard basis of $\mathbb{R}^3$.

Problem 2.3.11.
Suppose S is the rotation around the z-axis through $\pi/2$ counterclockwise (viewed from the positive z-axis) and T is the rotation around the x-axis through $\pi/2$ counterclockwise (viewed from the positive x axis).
a. **(S)** What is the axis of the rotation $T \circ S$?
b. **(C)** What is the angle of the rotation around the axis in (a)?

In computer graphics it is important to be able to give a geometric description of the composition of two rotations, but this can be very difficult. Quaternion techniques are often used for this purpose by programmers because of their relative efficiency. The following problems demonstrate one version of the technique. (Problems (2.2.3) and (2.2.4) provide some useful background.) Something of theoretical interest is involved here too. By (2.3.12) and (2.3.13) we have a function from S^3 onto $SO(\mathbb{R}^3)$. We will return to this function in 3.2, and in 5.4 we will see that this function is a special case of one of the important theorems in the Lie theory.

Problem 2.3.12. (C)
Let $q \in S^3$. By (2.2.4) we know $q = \cos(\alpha/2) + \sin(\alpha/2)u$ for some some $\alpha \in \mathbb{R}$ and some $u \in S^3$ with real part zero. Recall that the quaternions with real part zero can be identified with vectors in $\mathbb{R}^3$. Define a function $T_q \colon \mathbb{R}^3 \to \mathbb{R}^3$ by

$$T_q(v) = qvq^{-1} \quad \text{for } v \in \mathbb{R}^3.$$

a. Show that $T_q(v)$ has real part zero by calculating $T_q(v) + \overline{T_q(v)}$. This shows T_q really does map $\mathbb{R}^3$ to $\mathbb{R}^3$.
b. Show that T_q is a linear transformation: $T_q(v + w) = T_q(v) + T_q(w)$ and $T_q(rv) = rT_q(v)$ for all $v, w \in \mathbb{R}^3$ and all $r \in \mathbb{R}$.
c. Show that T_q preserves lengths: $T_q(v)\overline{T_q(v)} = v\overline{v}$ for all $v \in \mathbb{R}^3$, so $T_q \in O(\mathbb{R}^3)$.
d. Choose unit vectors v and w in $\mathbb{R}^3$ which are orthogonal to u and to each other and with $u = v \times w$ (cross product), so v, w, u is a "right-handed" basis for $\mathbb{R}^3$. Find the matrix of T_q with respect to the ordered basis v, w, u.
e. Explain why T_q is a rotation through an angle α around the axis spanned by u. Note that this means that the correspondence $q \mapsto T_q$ maps each element of S^3 to an element of $SO(\mathbb{R}^3)$.

Problem 2.3.13.
Continuing the notation of (2.3.12), define $F \colon S^3 \to SO(\mathbb{R}^3)$ by $F(q) = T_q$.
a. For which q in S^3 is it true that T_q is the identity function on $\mathbb{R}^3$?
b. Explain why F is onto.

Problem 2.3.14. (C)
Continue the notation of (2.3.12) and (2.3.13), and suppose T_ℓ is a rotation of $\mathbb{R}^3$ through an angle α_ℓ around an axis spanned by the unit vector u_ℓ for $\ell = 1, 2$. Let $q_\ell = \cos(\alpha_\ell/2) + \sin(\alpha_\ell/2)u_\ell \in \mathbb{H}$, so $T_\ell = F(q_\ell)$.

a. Find a formula for $q_2 q_1$ of the form $\cos(\beta/2) + \sin(\beta/2)u$ for $u \in S^3$ with real part zero.

b. Using your formula from (a), what is the axis of the rotation $T_2 \circ T_1$?

c. Using your formula from (a), what is the angle of the rotation $T_2 \circ T_1$?

d. Use quaternions to check your answers to (2.3.11).

2.4 Putting the pieces together

The **complex numbers** $\mathbb{C}$ extend the real numbers $\mathbb{R}$,

$$\mathbb{C} = \{a + bi : a, b \in \mathbb{R}\},$$

where the binary operations addition and multiplication are defined by $i^2 = -1$ and the usual rules in $\mathbb{R}$. That is,

$$(a + bi) + (a' + b'i) = (a + a') + (b + b')i,$$
$$(a + bi)(a' + b'i) = (aa' - bb') + (ab' + ba')i.$$

Another (unary) operation on $\mathbb{C}$ is **complex conjugation**, where the complex conjugate of $z = a + bi$ is

$$\overline{z} = a - bi.$$

A standard way to draw a picture of a complex number is to represent $a + bi$ by the vector $(a, b) \in \mathbb{R}^2$. In this representation, the horizontal axis is the *real* axis and the vertical is *imaginary*. The length of the vector corresponding to $z = a + bi$ is then $|z| = \sqrt{z\overline{z}}$. A complex number of length 1 has the form $e^{i\alpha} = \cos(\alpha) + i\sin(\alpha)$ for some real number α.

In a natural way, each complex number $e^{i\alpha}$ can be identified with a point of the unit circle (the 1-sphere) S^1 in $\mathbb{R}^2$. Conversely, each point of S^1 can be identified with a complex number of the form $e^{i\alpha}$:

$$S^1 = \{e^{i\alpha} : \alpha \in \mathbb{R}\} = \{z \in \mathbb{C} : |z| = 1\}.$$

We can extend complex conjugation to matrices and vectors by defining $\overline{A} = (\overline{a}_{ij})$ for a matrix A with complex entries a_{ij} and $\overline{\boldsymbol{v}} = (\overline{v}_1, \ldots, \overline{v}_n)$ for a vector $\boldsymbol{v}$ with complex entries v_j. The most important groups of matrices with complex entries are the **unitary group**

$$U(n, \mathbb{C}) = \{M \in \mathcal{M}(n, \mathbb{C}) : \overline{M}^T M = I_n\}$$

and its subgroup the **special unitary group**

$$SU(n, \mathbb{C}) = \{M \in U(n, \mathbb{C}) : \det(M) = 1\}.$$

In Chapter 6 we'll see that $U(n, \mathbb{C})$ corresponds to the group of isometries of a complex vector space with a "Hermitian" form which generalizes the Euclidean bilinear form.

The **quaternions** $\mathbb{H}$ are an extension of the complex numbers:

$$\mathbb{H} = \{a + bi + cj + dk : a, b, c, d \in \mathbb{R}\},$$

where $i^2 = j^2 = k^2 = -1$, $ij = k = -ji$, $jk = i = -kj$, and $ki = j = -ik$. Otherwise, addition and multiplication follow the usual rules for $\mathbb{R}$: for $z = a + bi + cj + dk$ and $z' = a' + b'i + c'j + d'k$,

$$\begin{aligned} z + z' &= (a + a') + (b + b')i + (c + c')j + (d + d')k, \\ zz' &= (aa' - bb' - cc' - dd') + (ab' + ba' + cd' - dc')i \\ &\quad +(ac' + ca' + da' - ad')j + (ad' + da' + bc' - cb')k. \end{aligned}$$

Addition and multiplication are associative and satisfy the distributive law. Addition is commutative, but multiplication is *not.* An analogue of complex conjugation is also defined for $\mathbb{H}$: $\overline{z} = a - bi - bj - bk$.

A quaternion $z = a+bi+cj+dk \in \mathbb{H}$ can be identified with the point $(a, b, c, d) \in \mathbb{R}^4$. A quaternion $z = bi + cj + dk$ with real part zero can be identified with $(b, c, d) \in \mathbb{R}^3$. If we define $|z| = \sqrt{z\overline{z}}$, then $|z|$ agrees with Euclidean length, whether in $\mathbb{R}^4$ or $\mathbb{R}^3$. The 3-sphere S^3 in $\mathbb{R}^4$ is identified with the quaternions of length 1:

$$S^3 = \{z \in \mathbb{H} \; : \; |z| = 1\}.$$

The 2-sphere in $\mathbb{R}^3$ is

$$S^2 = \{(x, y, z) \in \mathbb{R}^3 \; : \; x^2 + y^2 + z^2 = 1\}.$$

Unlike S^1 and S^3, S^2 does not have a natural group structure.

The proofs of the following theorems are embedded in the problems of this chapter.

Theorem 2.1 *Assume T is in $O(\mathbb{R}^3)$ and T has matrix M with respect to the standard basis. Then M has determinant 1 if and only if T is a rotation about a line through the origin. In the special case when the axis of the rotation T is the z-axis, there is some real number α for which*

$$M = \begin{bmatrix} \cos(\alpha) & -\sin(\alpha) & 0 \\ \sin(\alpha) & \cos(\alpha) & 0 \\ 0 & 0 & 1 \end{bmatrix}.$$

Theorem 2.2

a. *Complex conjugation satisfies $\overline{z_1 + z_2} = \overline{z}_1 + \overline{z}_2$, $\overline{z_1 z_2} = \overline{z}_1\,\overline{z}_2$. Further, $|z_1 z_2| = |z_1||z_2|$ for all $z_1, z_2 \in \mathbb{C}$.*

b. *The set $\mathbb{C}$ is an abelian group under addition, and the set of nonzero elements of $\mathbb{C}$ is an abelian group under multiplication. The 1-sphere S^1 is a subgroup of the latter group.*

Theorem 2.3 *$U(n, \mathbb{C})$ is a group under matrix multiplication, and $SU(n, \mathbb{C})$ is a subgroup. In the special case $n = 2$, every matrix in $SU(2, \mathbb{C})$ has the form*

$$M = \begin{bmatrix} u & -\overline{v} \\ v & \overline{u} \end{bmatrix}$$

for complex numbers u and v satisfying $u\overline{u} + v\overline{v} = 1$.

Theorem 2.4

a. *Quaternionic conjugation satisfies* $\overline{z_1 + z_2} = \overline{z}_1 + \overline{z}_2$ *and* $\overline{z_1 z_2} = \overline{z}_2\, \overline{z}_1$ *for all* $z_1, z_2 \in \mathbb{H}$. *Further* $|z_1 z_2| = |z_1||z_2|$ *for all* $z_1, z_2 \in \mathbb{H}$.

b. *The set* $\mathbb{H}$ *is an abelian group under addition, and the set of nonzero elements of* $\mathbb{H}$ *is a non-abelian group under multiplication. The 3-sphere* S^3 *is a non-abelian subgroup of the latter group.*

Theorem 2.5

a. *Every quaternion* q *in* S^3 *can be written in the form* $q = \cos(\alpha/2) + \sin(\alpha/2)u$ *for some* $\alpha \in \mathbb{R}$ *and some* $u \in S^3$ *with real part zero.*

b. *For* $q \in S^3$ *and* $v \in \mathbb{R}^3$, *the function* T_q *given by* $T_q(v) = qvq^{-1}$ *defines a rotation in* $SO(\mathbb{R}^3)$, *where we identify quaternions with real part zero with vectors in* $\mathbb{R}^3$. *In particular, if* $q = \cos(\alpha/2) + \sin(\alpha/2)u$ *as in (a), then* T_q *is a rotation through an angle* α *around an axis spanned by* u.

c. *The function* $F\colon S^3 \to SO(\mathbb{R}^3)$ *given by* $F(q) = T_q$ *is onto, but not one-to-one.*

Suggestions for further reading

Conway, J.H. and Smith, D.A., *On Quaternions and Octonions: Their Geometry, Arithmetic, and Symmetry*, A.K. Peters (2002). Chapter I on the complex numbers and the first two sections of chapter II on the quaternions are especially accessible.

Cromwell, P.R., *Polyhedra*, Cambridge University Press (1997). Conway and Smith discuss finite subgroups of $O(\mathbb{R}^3)$ and of $\mathbb{H}$ in their chapter III. Cromwell has a more accessible (and very lovely) treatment of the finite subgroups of $SO(\mathbb{R}^3)$ in his chapter 8 entitled "Symmetry, Shape and Structure."

2.5 A broader view: octonions

We've seen that the one-dimensional real numbers can be extended to the two-dimensional complex numbers. The arithmetic of the complex numbers has all of the nice properties of the arithmetic of the real numbers. The two-dimensional complex numbers can be extended to the four-dimensional quaternions. The arithmetic of the quaternions, however, is somewhat different: multiplication is not commutative. The four-dimensional quaternions can be extended to an eight-dimensional number system called the **octonions**. The octonions are described in terms of *seven* "imaginaries" which are often labelled $i, j, k, \ell, k\ell, j\ell$ and $i\ell$. The octonions $\mathbb{O}$ are then defined by

$$\mathbb{O} = \{a + bi + cj + dk + e\ell + fk\ell + gj\ell + hi\ell \,:\, a, b, c, d, e, f, g, h \in \mathbb{R}\}.$$

The addition of octonions is straightforward. To define multiplication, we begin by setting the square of each of these imaginaries equal to -1. We also need to say what the product of two different imaginaries is. The following diagram (adapted from Tevian Dray) shows how to multiply two imaginaries by following the arrows. For example, $(k\ell)(j) = i\ell$. Reversing the direction introduces a minus sign; for example, $(j)(k\ell) = -i\ell$.

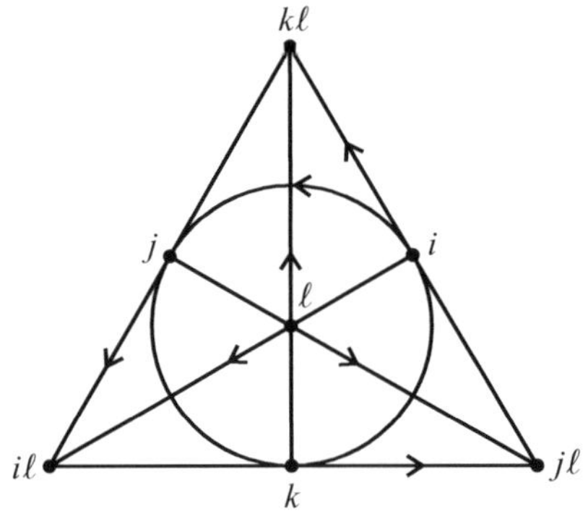

Figure 2.2. Multiplying imaginaries in the octonions (adapted from T. Dray)

These choices lead to the definition of the multiplication of two arbitrary octonions. Not only is this multiplication not commutative, it's not even associative!

In their book *On Quaternions and Octonians*, Conway and Smith quote from an account by the mathematical physicist John Baez of the discovery of the octonions. The day after Hamilton discovered the quaternions (October 16, 1843), he wrote about it to his college friend Graves.

> Graves replied on October 26th, complimenting Hamilton on the boldness of the idea, but adding, "There is still something in the system which gravels me. I have not yet any clear views as to the extent to which we are at liberty arbitrarily to create imaginaries, and to endow them with supernatural properties." And he asked: "If with your alchemy you can make three pounds of gold, why should you stop there?" Graves then set to work on some gold of his own! On December 26th, he wrote to Hamilton describing a new 8-dimensional algebra.

So now you may have two questions:

- Why should we care about octonions?
- Can we keep going, creating number systems of dimensions 16, 32, 64, and on and on indefinitely?

For an answer to the first question, we turn again to Baez, this time to a 2004 interview with him published in the online *Plus Magazine ... living mathematics* at `http://plus.maths.org/issue33/features/baez/index.html`. He said:

> I've spent a long time just trying to understand [the octonions]. What got me going on them is my interest in exceptions in mathematics. Often you can classify some sort of gizmo, and you get a beautiful systematic list, but also some number of exceptions. Nine times out of 10 those exceptions are related to the octonions. That intrigues me: what is the essence of octonions that makes them do what they do?

The interviewer then asked whether the octonions had anything to do with "reality". Baez answered:

> The most exciting example recently is in string theory, which is a theory of physics where people are trying to combine gravity with the other forces, and they found out that this theory only works if space and time have 10 dimensions. And the reason is

> that a string traces out a 2-dimensional surface in space and time, and it can wiggle around in whatever other directions are left available, and it turns out that the maths only works out nicely if there are 8 other directions for it to move around in!

The interviewer than writes, "Baez thinks that if string theory turned out to be true then octonions would become very fashionable."

What about the other question? Can we keep going, creating higher and higher dimensional number systems? Here an answer is via an 1898 theorem of the German mathematician Adolf Hurwitz. The four number systems $\mathbb{R}$, $\mathbb{Q}$, $\mathbb{H}$ and $\mathbb{O}$ are what Conway and Smith call "composition algebras with 1". (See their chapter 6.) That means each is a vector space over the real numbers with a multiplication rule and having a multiplicative identity 1, and further, each has a notion of length given by $|z|^2 = \overline{z}z$ satisfying $|z_1z_2| = |z_1||z_2|$. Hurwitz proved that the only composition algebras with 1 are these four. After the octonions, there are no more!

CHAPTER 3

Linearization

3.1 Tangent spaces

In calculus, a line tangent to the graph of a differentiable function $\mathbb{R} \to \mathbb{R}$ is a good approximation to the actual graph near the point of tangency. Linearization replaces a possibly complicated function by a linear function, one that is much simpler and yet still captures some information about the original. Similarly, the graph of a (smoothly) differentiable function $\mathbb{R}^2 \to \mathbb{R}$ is a smooth surface in $\mathbb{R}^3$. A plane tangent to the surface is a good approximation to the actual graph near the point of tangency. Again, linearization replaces a possibly complicated function by a linear function. In this section, we will define a tangent space for a matrix group. A matrix group typically is very complicated, but its tangent space is a vector space, a much simpler mathematical structure.

In fact, it's a little *too* simple. As we'll see in Chapter 5, by adding just a little more structure to the tangent space, we will turn it into a *Lie algebra*, and this Lie algebra will encode important information about the original group, the Lie group. Indeed, the translation of questions about Lie groups into questions about Lie algebras lies at the heart of the Lie theory. In this chapter, however, we're just concerned with the tangent space as a vector space.

To motivate our definition of the tangent space to a matrix group, we'll think in a somewhat different way about the plane tangent to a smooth surface $\mathcal{S}$ in $\mathbb{R}^3$ at a point $\boldsymbol{p}$ on the surface. First, we need to be more precise about what we mean by "smooth." A *smooth surface* in $\mathbb{R}^3$ is one that has a tangent plane at every point. More generally, a function $f: \mathbb{R} \to \mathbb{R}$ is **smooth** if f is *infinitely* differentiable—the nth derivative of f exists for all positive integers n. Similarly, a function $f: \mathbb{R} \to \mathbb{R}^N$, with $f(x) = (f_1(x), \ldots, f_N(x))$ is smooth if all of the component functions f_j are smooth. And finally, a function $F: \mathbb{R}^M \to \mathbb{R}^N$ with $F(x_1, \ldots, x_M) = (f_1(x_1, \ldots, x_M), \ldots, f_N(x_1, \ldots, x_M))$ is smooth if all of the component functions f_j are smooth functions of each x_i; that is, if partial derivatives of f_j of all orders exist. (We'll see some specific calculus examples in section 3.)

So, back to our smooth surface $\mathcal{S}$ in $\mathbb{R}^3$. Imagine a smooth curve drawn on the surface passing through $\boldsymbol{p}$. More formally, assume the curve is parameterized as the image of a smooth function $\gamma: \mathbb{R} \to \mathcal{S}$, with, say, $\gamma(t_0) = \boldsymbol{p}$. (Usually, t_0 will be 0.) Write $\gamma(t) = (x(t), y(t), z(t))$. Recall from multivariable calculus that $\gamma'(t) = (x'(t), y'(t), z'(t))$, and we can interpret $\gamma'(t_0)$ as a vector tangent to the curve at $\boldsymbol{p}$. Since the curve lies on

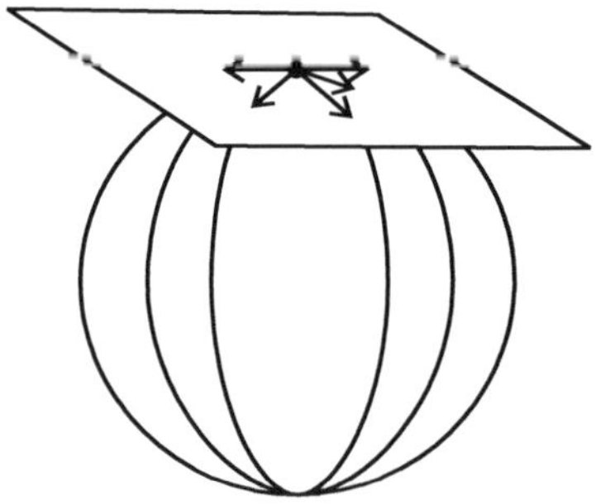

Figure 3.1. Sphere with "north pole" $\boldsymbol{p}$ and tangent plane at $\boldsymbol{p}$

the surface, $\gamma'(t_0)$ is also tangent to $\mathcal{S}$ at $\boldsymbol{p}$. Now imagine many smooth curves passing through $\boldsymbol{p}$ on $\mathcal{S}$. Each of the curves will have a tangent vector at $\boldsymbol{p}$. All these tangent vectors will lie in the plane tangent to $\mathcal{S}$ at $\boldsymbol{p}$. In fact, these tangent vectors determine the plane.

Here's a relatively simple example to hold in your mind's eye. Imagine the surface $\mathcal{S}$ is a sphere and the point $\boldsymbol{p}$ is its "north pole." Consider the great circles passing through $\boldsymbol{p}$. These are all smooth curves on $\mathcal{S}$. For each parameterization of a particular great circle, imagine its tangent vector at $\boldsymbol{p}$. Visualize these tangent vectors "fanning out" in the plane tangent to the sphere at $\boldsymbol{p}$. (See Figure 3.1.)

This is the strategy we imitate to define a tangent space to a matrix group $G \subseteq GL(n, \mathbb{R})$. We'll think of n by n matrices as if they were N-tuples for $N = n^2$, so $G \subseteq \mathbb{R}^N$. The group G is like the surface $\mathcal{S}$. The special "point" on the "surface" G is $I_n \in G$. We will consider smooth curves on G which pass through I_n.

The tangent space at the identity of G is the set of all matrices having the form $\gamma'(0)$, for some smooth function $\gamma: \mathbb{R} \to G$ which satisfies $\gamma(0) = I_n$.

We set up our notation so that the n by n matrix $\gamma(t)$ has i, j entry $a_{ij}(t)$ with $a_{ij}(t)$ a smooth function. Then $\gamma'(t)$ is an n by n matrix with i, j entry $a'_{ij}(t)$. The image of γ is called a **smooth curve** in G **passing through the identity**, and $\gamma'(0)$ is a **tangent vector** to that curve at the identity. In honor of Lie, the **tangent space at the identity** of G is denoted by $L(G)$.

$$L(G) = \{\gamma'(0) \mid \gamma: \mathbb{R} \to G,\ \gamma(0) = I_n,\ \gamma \text{ smooth}\}.$$

First we look at some examples of specific tangent vectors. In the problems that follow, all the coordinate functions $a_{ij}(t)$ are familiar from calculus and known to be smooth functions of t, so the corresponding curve can be assumed smooth. You will have to check, however, that the image of the curve lies in the appropriate group and passes through the identity of that group.

Problem 3.1.1.

Let $t \in \mathbb{R}$. Define a function γ by

$$\gamma(t) = \begin{bmatrix} \cosh(t) & \sinh(t) \\ \sinh(t) & \cosh(t) \end{bmatrix}.$$

a. Show that γ gives a (smooth) curve in the Lorentz group $\mathcal{L}(2, \mathbb{R})$ passing through the identity. That is, show $\gamma(t) \in \mathcal{L}(2, \mathbb{R})$ and $\gamma(0) = I_2$.

b. Find $\gamma'(0)$.

Problem 3.1.2.

Let $t \in \mathbb{R}$. Define a function γ by

$$\gamma(t) = \begin{bmatrix} \cos(t) & -\sin(t) & 0 \\ \sin(t) & \cos(t) & 0 \\ 0 & 0 & 1 \end{bmatrix}.$$

a. Show that γ gives a (smooth) curve in the Euclidean group $O(3, \mathbb{R})$ passing through the identity.

b. Find $\gamma'(0)$.

Problem 3.1.3. (S)

a. With $\gamma(t)$ as in (3.1.2), find the matrix $\sigma(t) = (\gamma(t))^2$, and use trig identities to simplify the entries.

b. Show that σ gives a (smooth) curve in the Euclidean group $O(3, \mathbb{R})$ passing through the identity.

c. What is the relationship between $\sigma'(0)$ and $\gamma'(0)$?

Problem 3.1.4.

Using problem (3.1.2) for inspiration, find a (smooth) curve in the Euclidean group $O(3, \mathbb{R})$ passing through the identity whose tangent vector at the identity is

$$\begin{bmatrix} 0 & 0 & 0 \\ 0 & 0 & -1 \\ 0 & 1 & 0 \end{bmatrix}.$$

Now we return to the general case. Fix a positive integer n and assume G is a subgroup of $GL(n, \mathbb{R})$, so the elements of $L(G)$ belong to $\mathcal{M}(n, \mathbb{R})$. Recall that $\mathcal{M}(n, \mathbb{R})$ is itself a vector space over the real numbers. The "vectors" are the matrices. Vector addition is the ordinary addition of matrices: if matrix A has entries a_{ij} and matrix B has entries b_{ij}, then $A + B$ has entries $a_{ij} + b_{ij}$. This is exactly vector addition in $\mathbb{R}^N$ for $N = n^2$. Scalar multiplication is also defined as in $\mathbb{R}^N$: if A is an n by n matrix with entries a_{ij} and c is a real number, then the matrix cA has entries ca_{ij}. In (3.1.5)–(3.1.8) you will show that if G is a subgroup of $GL(n, \mathbb{R})$, then $L(G)$ is a *subspace* of the vector space $\mathcal{M}(n, \mathbb{R})$.

Problem 3.1.5.

Fix a positive integer n and assume G is a group of n by n matrices. Show that the zero matrix $\mathbf{0}_n$ is always in $L(G)$. In other words, make up a function $\gamma: \mathbb{R} \to G$ giving a (smooth) curve passing through the identity and satisfying $\gamma'(0) = \mathbf{0}_n$. (Note: The group G has not been specified, but that does no harm. What matrix can you be sure is in *every* group of n by n matrices?)

Problem 3.1.6.

In this problem, you'll show that $L(G)$ is closed under scalar multiplication. Part (a) is an example, and parts (b) and (c) give the general argument.

a. Choose $G = SL(2, \mathbb{R})$, let $t \in \mathbb{R}$ and let

$$\alpha(t) = \begin{bmatrix} e^t & 0 \\ 0 & e^{-t} \end{bmatrix}.$$

Let $\gamma(t) = \alpha(17t)$. Verify that each of α and γ gives a (smooth) curve in G passing through the identity. Find $\alpha'(0)$ and $\gamma'(0)$. What is the relationship between these two vectors?

b. Now go back to the general case, fix $n > 0$, and assume G is a subgroup of $GL(n, \mathbb{R})$. Suppose A is in $L(G)$; that is, assume $A = \alpha'(0)$ for a (smooth) curve α passing through the identity of G. Choose an arbitrary real number c. To show that the matrix cA is also in $L(G)$, consider the function γ defined by

$$\gamma(t) = \alpha(ct).$$

First, explain why γ gives a (smooth) curve in G passing through the identity. Then show that $\gamma'(0) = cA$. (Hint: Assume the ij entry of $\alpha(t)$ is $a_{ij}(t)$, and use the chain rule.)

c. Why does showing $\gamma'(0) = cA$ in (b) tell you that cA is in $L(G)$?

Problem 3.1.7.

Show that the product rule holds for derivatives of matrix-valued functions in the special case of 2 by 2 matrices. That is, let

$$\alpha(t) = \begin{bmatrix} a_{11}(t) & a_{12}(t) \\ a_{21}(t) & a_{22}(t) \end{bmatrix} \quad \text{and} \quad \beta(t) = \begin{bmatrix} b_{11}(t) & b_{12}(t) \\ b_{21}(t) & b_{22}(t) \end{bmatrix}.$$

Check that $[\alpha(t)\beta(t)]' = \alpha(t)\beta'(t) + \alpha'(t)\beta(t)$ for all $t \in \mathbb{R}$.

Conceptually, the proof of the product rule for n by n matrices is no different than in the 2 by 2 case, and from now on we'll assume the **product rule** holds for differentiable functions $\alpha, \beta \colon \mathbb{R} \to GL(n, \mathbb{R})$ for any $n > 0$:

$$[\alpha(t)\beta(t)]' = \alpha(t)\beta'(t) + \alpha'(t)\beta(t)$$

for all $t \in \mathbb{R}$. (Notice that, as in the calculation in (3.1.7), the invertibility of the matrices in $GL(n, \mathbb{R})$ plays no role here.)

Problem 3.1.8. (S)

In this problem, you'll show that $L(G)$ is closed under addition. Part (a) is an example, and parts (b) and (c) give the general argument.

a. Choose $G = SL(2, \mathbb{R})$, let $\alpha(t)$ be as in (3.1.6a), and let

$$\beta(t) = \begin{bmatrix} 1 & t \\ 0 & 1 \end{bmatrix}.$$

Let $\gamma(t) = \alpha(t)\beta(t)$. Verify that each of β and γ gives a (smooth) curve in G passing through the identity. Find $\beta'(0)$ and $\gamma'(0)$. How is the vector $\gamma'(0)$ related to the vectors $\alpha'(0)$ and $\beta'(0)$?

b. Now go back to the general case, fix $n > 0$, and assume G is a subgroup of $GL(n, \mathbb{R})$. Assume that $A = \alpha'(0)$ and $B = \beta'(0)$ are two elements of $L(G)$, for α, β smooth curves in G passing through the identity. To show that the matrix $A + B$ is also in $L(G)$, consider the function γ defined by $\gamma(t) = \alpha(t)\beta(t)$. First explain why γ gives a (smooth) curve in G passing through the identity. Then show that $\gamma'(0) = A + B$.

c. Why does showing $\gamma'(0) = A + B$ in (b) tell you that $A + B$ is in $L(G)$?

We will use the content of the following linear algebra problem over and over again to prove that two vector spaces are equal. You may cite the result of (3.1.9) in subsequent problems.

Problem 3.1.9. (S)

Suppose U and W are both subspaces of some ambient vector space, say $\mathcal{M}(n, \mathbb{R})$. (Recall that $\mathcal{M}(n, \mathbb{R})$ is a subspace of itself, although not a *proper* subspace.) Assume that U is also a subset of W, $U \subseteq W$. Suppose that $\boldsymbol{w}_1, \boldsymbol{w}_2, \ldots, \boldsymbol{w}_m$ is a basis for W and also that $\boldsymbol{w}_j \in U$ for $j = 1, 2, \ldots, m$. Show that *every* vector in W is contained in U; that is, $W \subseteq U$. (Note that it follows from $U \subseteq W$ and $W \subseteq U$ that $U = W$.)

In the problems that follow, you should *assume* that if G is a subgroup of $GL(n, \mathbb{R})$, then $L(G)$ is a subspace of the vector space $\mathcal{M}(n, \mathbb{R})$.

Problem 3.1.10.

In this problem you will determine $L(G)$ for $G = GL(2, \mathbb{R})$.

a. Check that the following function gives a (smooth) curve in G passing through the identity.

$$\gamma_1(t) = \begin{bmatrix} e^t & 0 \\ 0 & 1 \end{bmatrix}.$$

Find the corresponding tangent vector $\gamma_1'(0) \in L(G)$.

b. Check that the following function also gives a (smooth) curve in G passing through the identity.

$$\gamma_2(t) = \begin{bmatrix} 1 & t \\ 0 & 1 \end{bmatrix}.$$

Find the corresponding tangent vector $\gamma_2'(0) \in L(G)$.

c. Find *two more* tangent vectors $\gamma_3'(0)$ and $\gamma_4'(0)$ in $L(G)$ so that $\gamma_1'(0)$, $\gamma_2'(0)$, $\gamma_3'(0)$ and $\gamma_4'(0)$ are linearly independent. Do this by producing two more functions γ_3 and γ_4 giving (smooth) curves in G passing through the identity.

d. Explain why your four tangent vectors form a basis for the vector space $\mathcal{M}(2, \mathbb{R})$.

e. Explain why $L(G) = \mathcal{M}(2, \mathbb{R})$ in this case. (Use (3.1.9), making clear what plays the role of U and what plays the role of W in this case.)

Problem 3.1.11. (S)

In this problem you will find $L(G)$ for $G = SL(2, \mathbb{R})$.

a. Suppose that $\gamma(t)$ gives a (smooth) curve in G passing through the identity, with

$$\gamma(t) = \begin{bmatrix} a(t) & b(t) \\ c(t) & d(t) \end{bmatrix}.$$

What are the values of $a(0), b(0), c(0)$ and $d(0)$?

b. Use the fact that the determinant of $\gamma(t)$ equals 1 to write an equation involving $a(t), b(t), c(t)$ and $d(t)$. Differentiate both sides of this equation with respect to t and evaluate at $t = 0$. What does this tell you about the form of a "typical" vector $\gamma'(0)$ in $L(G)$?

c. Your conclusion in (b) should tell you that $a'(0) + d'(0) = 0$, so $L(G)$ is a subset of

$$W = \left\{ \begin{bmatrix} a & b \\ c & -a \end{bmatrix} : a, b, c \in \mathbb{R} \right\}.$$

(i) Verify that W is a subspace of $\mathcal{M}(2, \mathbb{R})$ by checking that W contains the zero matrix $\mathbf{0}_2$ and is closed under addition and scalar multiplication.

(ii) Find a basis for W.

d. Show that $L(G)$ contains each basis vector of W. (Remember, to show a specific vector is in $L(G)$, you must find a (smooth) curve $\gamma(t)$ in G passing through the identity with $\gamma'(0)$ equal to the desired vector.)

e. Explain why $L(G) = W$.

Problem 3.1.12.

In this problem you will find $L(G)$ for $G = O(2, \mathbb{R})$ in the Euclidean case.

a. Suppose $\gamma(t)$ is a (smooth) curve in G passing through the identity. We know a matrix $\gamma(t)$ is in G if and only if $\gamma(t)^T \gamma(t) = I_2$, where $\gamma(t)^T$ denotes the transpose of $\gamma(t)$. Differentiate both sides of this matrix equation, using the product rule for matrices.

b. Show that $(\gamma(t)^T)' = \gamma'(t)^T$.

c. Combine the results of (a) and (b) and evaluate at $t = 0$. What does this tell you about the form of a "typical" vector $\gamma'(0)$ in $L(G)$?

d. Based on your conclusion in (c), $L(G)$ is a subspace of a subspace W of $\mathcal{M}(2, \mathbb{R})$.

(i) What is W?

(ii) Verify that W really is a subspace.

(iii) Find a basis of W.

e. Show that $L(G)$ contains the basis vectors of W.

f. Explain why $L(G) = W$.

Problem 3.1.13.

Let $G = \mathcal{L}(2, \mathbb{R})$. Find $L(G)$.

Problem 3.1.14.

Suppose $H = SO(2, \mathbb{R})$, the group of (Euclidean) rotations and $G = O(2, \mathbb{R})$. Show that $L(H) = L(G)$. (Cite the results in (3.1.12) wherever needed; don't re-prove them.)

Problem 3.1.15.

Let $G = GL(n, \mathbb{R})$. Show $L(G) = \mathcal{M}(n, \mathbb{R})$.

Problem 3.1.16.
For this problem you can assume $(\gamma(t)^T)' = \gamma'(t)^T$ holds for n by n matrices.

a. Go as far as you can in determining $L(G)$ in the general case

$$G = O(n, \mathbb{R}) = \{M \in \mathcal{M}(n, \mathbb{R}) \mid M^T C M = C\},$$

for a fixed choice of a symmetric matrix $C \in GL(n, \mathbb{R})$. (We'll complete the argument in the next chapter, making use of the matrix exponential.)

b. If instead of being symmetric, $C^T = -C$, then G would be the symplectic group $Sp\,(n, \mathbb{R})$. Does your argument in (a) apply to this case? Are any changes needed?

3.2 Group homomorphisms

In this section we broaden our repertoire of symmetries of mathematical structures. As we've already seen, a symmetry of a vector space—preserving the vector space operations of addition and scalar multiplication—is an invertible linear transformation. What about a symmetry of a group? The special feature of a group is its binary operation. The operation is matrix multiplication if the group elements are matrices, composition if the group elements are functions, multiplication if the group elements are nonzero complex numbers, etc. To be general and include all these possibilities as well as others, we'll denote the binary operation by the symbol $*$. A symmetry of a group G with binary operation $*$ is an invertible function $F\colon G \to G$ that preserves the binary operation:

$$F(g * h) = F(g) * F(h)$$

for all $g, h \in G$. More generally, a **group isomorphism** from a group G_1 with binary operation $*$ to a group G_2 with binary operation $\diamond$ is a one-to-one and onto function $F\colon G_1 \to G_2$ satisfying

$$F(g * h) = F(g) \diamond F(h) \tag{3.1}$$

for all $g, h \in G_1$. We also say that groups G_1 and G_2 are **isomorphic** if there is an isomorphism from one to the other. Notice that the word "isomorphic" doesn't indicate which group is the domain of the isomorphism and which is the range. That's because if $F\colon G_1 \to G_2$ is an isomorphism, then $F^{-1}\colon G_2 \to G_1$ is also, as you'll show in (3.2.12).

When the groups G_1 and G_2 are isomorphic, we write $G_1 \simeq G_2$. The resemblance between the symbol $\simeq$ and an equal sign is no accident; two isomorphic groups can be regarded as essentially "the same"—we can think of the one-to-one correspondence F as merely assigning new names to the elements of G_1. This is exactly the situation with our groups of linear transformations and groups of matrices in Chapter 1; see (3.2.4) below.

Recall that analogous language is used for vector spaces and linear transformations. Suppose V_1 and V_2 are vector spaces (over the real numbers) and $T\colon V_1 \to V_2$ is a linear transformation. If T is one-to-one and onto (i.e., if T is invertible), then T is called a **vector space isomorphism**, and we say V_1 and V_2 are **isomorphic vector spaces**. Just as it's useful to relax the requirement of invertibility and consider linear transformations, it's also useful to relax the requirement that the function F be one-to-one and onto, and simply require that it preserve the binary operation as in equation (3.1). Such a function is called a **group homomorphism**.

Problem 3.2.1.

As a warm-up, here is a vector space example. Choose the vector spaces

$$V_1 = \left\{ \begin{bmatrix} 0 & -b \\ b & 0 \end{bmatrix} : b \in \mathbb{R} \right\} \quad \text{and} \quad V_2 = \left\{ \begin{bmatrix} 0 & b \\ b & 0 \end{bmatrix} : b \in \mathbb{R} \right\}.$$

a. Verify that V_1 and V_2 are vector spaces over $\mathbb{R}$.

b. Define $T\colon V_1 \to V_2$ by

$$T\left(\begin{bmatrix} 0 & -b \\ b & 0 \end{bmatrix}\right) = \begin{bmatrix} 0 & b \\ b & 0 \end{bmatrix}.$$

Show T is a vector space isomorphism by verifying that T is a linear transformation and T is one-to-one and onto.

Problem 3.2.2.

a. Let G_1 be the set of real numbers. Verify that G_1 is a group under addition.

b. Let G_2 be the set of positive real numbers. Verify that G_2 is a group under multiplication.

c. Define $F(x) = e^x$. Verify that F is a group isomorphism from G_1 to G_2. (That is, check that F does map real numbers to positive real numbers, F preserves the two operations in the sense of equation (3.1), and further, that F is one-to-one and onto.)

d. Let G_3 be the set of *all* nonzero real numbers. You can assume that G_3 is a group under multiplication; the proof is essentially the same as for G_2. Let $G_4 = GL(1, \mathbb{R})$, the group of invertible 1 by 1 matrices under matrix multiplication. (Assume $[a][b] = [ab]$ and $\det([a]) = a$.) Show that $G_3 \simeq G_4$.

Problem 3.2.3.

Fix $n > 0$ and suppose $G_1 = GL(n, \mathbb{R})$ under matrix multiplication, and G_2 is the group of nonzero real numbers under multiplication. For $A \in G_1$, define $F(A) = \det(A)$.

a. Check that F is a group homomorphism from G_1 to G_2. (That is, check that F does take elements of G_1 to elements of G_2 and that equation (3.1) is satisfied.)

b. Assume $n = 2$. Show that F is an onto function: for every nonzero real number a in G_2, there is a matrix $A \in G_1$ with $F(A) = a$. In other words, make up a matrix A that does the job. (Your A will depend on a, of course.)

c. Still for the case $n = 2$, show that F is *not* one-to-one by finding two different matrices $A, B \in GL(2, \mathbb{R})$ with $F(A) = F(B)$. So, F is not an isomorphism from G_1 to G_2.

In the next problem, you'll justify the claims in Chapter 1 that the groups $GL(\mathbb{R}^n)$ and $GL(n, \mathbb{R})$ are "the same," and that the Euclidean groups $O(\mathbb{R}^n)$ and $O(n, \mathbb{R})$ are "the same." Similar arguments apply to the other corresponding pairs of a group of linear transformations and a group of matrices.

Problem 3.2.4.

In this problem, cite results from Chapter 1 (or from linear algebra) as needed; don't re-prove them.

a. Define a function $F: GL(\mathbb{R}^n) \to GL(n, \mathbb{R})$ by $F(T) = M$, where M is the matrix of T with respect to the standard basis of $\mathbb{R}^n$. Explain why F is a group isomorphism.

b. Assume both the transformation and the matrix orthogonal groups are defined with respect to the same non-degenerate symmetric bilinear form. Explain why restricting the function F defined in (a) to elements of $O(\mathbb{R}^n)$ gives a group isomorphism from $O(\mathbb{R}^n)$ to $O(n, \mathbb{R})$.

Problem 3.2.5.

Recall that $S^1 = \{e^{i\alpha} : \alpha \in \mathbb{R}\}$ is a group under multiplication of complex numbers. As usual, $\mathbb{R}$ is a group under addition. Define $F: \mathbb{R} \to S^1$ by $F(\alpha) = e^{i\alpha}$.

a. Show that F is a group homomorphism.

b. Show that F is onto.

c. Show that F is *not* one-to-one; specifically, describe the set Γ of real numbers α which satisfy $F(\alpha) = 1$. (If you've studied abstract algebra, you recognize that Γ is the kernel of F, and by the first isomorphism theorem $S^1 \simeq \mathbb{R}/\Gamma$. Notice that Γ is discrete—there are "spaces" between its elements. This is a special case of a theorem we'll discuss in 5.4. By the way, Γ is the upper case form of the Greek letter γ and is called gamma.)

Problem 3.2.6.

Define $F: SL(n, \mathbb{R}) \to SL(n, \mathbb{R})$ by $F(X) = (X^{-1})^T$.

a. Verify that if X is in $SL(n, \mathbb{R})$, then $F(X)$ is also in $SL(n, \mathbb{R})$.

b. Verify that for X, Y in $GL(n, \mathbb{R})$,

$$(XY)^{-1} = Y^{-1}X^{-1}$$

by calculating $(XY)(Y^{-1}X^{-1})$. (Remember that, in general, matrix multiplication is not commutative, so the order of the factors matters.)

c. Verify that F is a group homomorphism; i.e., show that $F(XY) = F(X)F(Y)$.

d. Verify that F is one-to-one and onto.

e. Find a formula for F^{-1} (using "formula" in the same sense in which the formula for F was specified).

Problem 3.2.7. (S)

(Compare the results of this problem to the example at the end of 1.6.) Choose $G_1 = \mathbb{Q}$ with group operation addition. Choose

$$G_2 = \left\{ \begin{bmatrix} 1 & a \\ 0 & 1 \end{bmatrix} : a \in \mathbb{Q} \right\}.$$

Define $F: G_1 \to G_2$ by

$$F(a) = \begin{bmatrix} 1 & a \\ 0 & 1 \end{bmatrix}.$$

a. Verify that G_1 is a group under addition.
b. Verify that G_2 is a group under matrix multiplication.
c. Verify that F is a group isomorphism.
d. Find a formula for F^{-1} and use it to verify that F^{-1} is also a group isomorphism.

Problem 3.2.8.
Choose $G_1 = SO(2, \mathbb{R})$, the matrix form of the group of Euclidean rotations of the plane and

$$G_2 = S^1 = \{e^{i\alpha} : \alpha \in \mathbb{R}\},$$

a group under multiplication of complex numbers. Define $F: G_1 \to G_2$ by

$$F\left(\begin{bmatrix} \cos(\alpha) & -\sin(\alpha) \\ \sin(\alpha) & \cos(\alpha) \end{bmatrix}\right) = e^{i\alpha}.$$

Verify that F is a group isomorphism.

A special property of a group homomorphism is that it maps the identity in the domain group to the identity in the range group. The analogous fact is true for a linear transformation: the zero vector in the domain space is mapped to the zero vector in the range space—you proved the special case when both spaces are $\mathbb{R}^n$ in (1.1.10), but the same argument generalizes. Problem (3.2.9) below provides illustrative examples for group homomorphisms, and (3.2.10) and (3.2.11) ask for proofs, first for matrix groups and then for arbitrary groups.

Problem 3.2.9.
Verify this property for the homomorphism F in
a. (3.2.2).
b. (3.2.3).
c. (3.2.4).
d. (3.2.5).
e. **(S)** (3.2.6).
f. **(S)** (3.2.7).
g. (3.2.8).

Problem 3.2.10.
Fix positive integers n and m. In this problem you'll prove that if G_1 and G_2 are matrix groups (with $G_1 \subseteq GL(n, \mathbb{R})$ and $G_2 \subseteq GL(m, \mathbb{R})$), and if $F: G_1 \to G_2$ is a group homomorphim, then $F(I_n) = I_m$.
a. Suppose $M \in GL(m, \mathbb{R})$ and $MM = M$. Show $M = I_m$.
b. Let $M = F(I_n)$ and then apply (a).

Problem 3.2.11.
Suppose G_1 and G_2 are groups with operations $*$ and $\diamond$ respectively. Suppose E_1 is the identity in G_1 and E_2 is the identity in G_2. Suppose $F: G_1 \to G_2$ is a group homomorphism. Show $F(E_1) = E_2$.

Problem 3.2.12.

Assume that G_1 and G_2 are groups with operations $*$ and $\diamond$ respectively, and assume that $F: G_1 \to G_2$ is a group isomorphism. Because F is one-to-one and onto, we know it is invertible; that is, it has an inverse $F^{-1}: G_2 \to G_1$ which is also one-to-one and onto. Show that F^{-1} is a group homomorphism. (Hint: set up notation with $F(X_1) = X_2, F(Y_1) = Y_2$. What is $F^{-1}(X_2)$? $F^{-1}(Y_2)$?)

Problem 3.2.13.

This problem shows that $S^3 = \{q \in \mathbb{H} : q\overline{q} = 1\}$ and $SU(2, \mathbb{C})$ are isomorphic groups. Define $F: S^3 \to SU(2, \mathbb{C})$ by

$$F(q) = F(a + bi + cj + dk) = \begin{bmatrix} a + bi & -c + di \\ c + di & a - bi \end{bmatrix}.$$

a. Show $q \in S^3$ implies $F(q) \in SU(2, \mathbb{C})$.
b. Show F is a group isomorphism.

Problem 3.2.14.

Recall from (2.3.12) and (2.3.13) the map from $S^3 = \{q \in \mathbb{H} : q\overline{q} = 1\}$ onto the Euclidean group $SO(\mathbb{R}^3)$ given by $q \mapsto T_q$, where $T_q(v) = qvq^{-1}$ for v a quaternion with real part zero (so v can be identified with a vector in $\mathbb{R}^3$). Write $F(q) = T_q$. Show that F is a group homomorphism, $F: S^3 \to SO(\mathbb{R}^3)$.

3.3 Differentials

Suppose we have a homomorphism $F: G_1 \to G_2$ from one Lie group (i.e., matrix group) to another. We seek a corresponding linear transformation from $L(G_1)$ to $L(G_2)$. The differential of F, denoted by dF, is the linear transformation we're looking for. As for tangent spaces, we motivate the definition by an example from calculus.

The calculus example is a tale of *two* surfaces, so we'll use subscripts to keep things straight. Suppose $\mathcal{S}_1$ is a smooth surface in $\mathbb{R}^3$, so, in particular, $\mathcal{S}_1$ has a tangent plane at the point $\boldsymbol{p}_1$. Suppose further that $\gamma_1: \mathbb{R} \to \mathcal{S}_1$ is a smooth curve passing through the point $\boldsymbol{p}_1$ on $\mathcal{S}_1$; to be definite, say $\gamma_1(t_0) = \boldsymbol{p}_1$. Now suppose $\mathcal{S}_2$ is another smooth surface in $\mathbb{R}^3$. Suppose further that we have a smooth function F mapping points on the first surface to points on the second, $F: \mathcal{S}_1 \to \mathcal{S}_2$. Since $\mathcal{S}_2$ is smooth, it has a tangent plane at $\boldsymbol{p}_2 = F(\boldsymbol{p}_1)$.

Recall what it means to say the multivariable function F is smooth. The function F takes triples to triples, so we can write

$$F(x, y, z) = (f_1(x, y, z), f_2(x, y, z), f_3(x, y, z))$$

where each of the f_j maps points in $\mathbb{R}^3$ to real numbers. To say that F is smooth means that each of the f_j is smooth, and that means, in particular, that the partial derivatives $\partial f_j/\partial x$, $\partial f_j/\partial y$, and $\partial f_j/\partial z$ all exist for $j = 1, 2, 3$.

Imagine that we apply F to each of the points of the curve determined by $\gamma_1(t)$. The resulting points lie along a curve in $\mathcal{S}_2$, and the image curve passes through the point

$\boldsymbol{p}_2 = F(\boldsymbol{p}_1)$ on the surface $\mathcal{S}_2$. More formally, let $\gamma_2(t) = F(\gamma_1(t))$. Then $\gamma_2 \colon \mathbb{R} \to \mathcal{S}_2$ and $\gamma_2(t_0) = F(\gamma_1(t_0)) = F(\boldsymbol{p}_1) = \boldsymbol{p}_2$. Is γ_2 smooth? Yes, by the chain rule, since F and γ_1 are smooth.

Problem 3.3.1. (S)
Here's a specific example with surfaces and curves you should be able to visualize—the parts of this problem asking for descriptions are asking for descriptions of what the object *looks* like or, even better, a sketch. Let $\mathcal{S}_1$ be the cone

$$\mathcal{S}_1 = \{(x, y, z) : z = \sqrt{x^2 + y^2}\}.$$

Note that $\mathcal{S}_1$ actually isn't smooth at the origin (why?), but we'll avoid that point, choosing $\boldsymbol{p}_1 = (3, 0, 3)$. Define $F(x, y, z) = (y, x, z^2)$.

a. Describe $\mathcal{S}_2 = F(\mathcal{S}_1)$.

b. Let $\gamma_1(t) = (t, 0, t)$ for $t \geq 0$. Explain why $\gamma_1(t) \in \mathcal{S}_1$ and describe the curve on $\mathcal{S}_1$ parameterized by γ_1. We want $\gamma_1(t_0) = (3, 0, 3)$, so what is t_0? $\gamma_1'(t_0)$? Describe $\gamma_1'(t_0)$.

c. What is the formula for $\gamma_2(t) = F(\gamma_1(t))$? Describe the curve on $\mathcal{S}_2$ parameterized by γ_2. What is $\gamma_2(t_0)$? $\gamma_2'(t_0)$? Describe $\gamma_2'(t_0)$.

Returning to the general situation where $\mathcal{S}_1$ is a smooth surface in $\mathbb{R}^3$, γ_1 is a smooth curve on $\mathcal{S}_1$, and F is a smooth function $\mathbb{R}^3 \to \mathbb{R}^3$, consider the tangent vector $\gamma_1'(t_0)$ in the plane tangent to $\mathcal{S}_1$ at $\boldsymbol{p}_1 = \gamma_1(t_0)$. There is a corresponding tangent vector $\gamma_2'(t_0)$ in the plane tangent to $\mathcal{S}_2$ at $\boldsymbol{p}_2 = F(\boldsymbol{p}_1)$. In other words, the function F not only maps points of $\mathcal{S}_1$ to points of $\mathcal{S}_2$, it also determines a function that maps vectors tangent to curves on $\mathcal{S}_1$ passing through $\boldsymbol{p}_1$ to vectors tangent to curves on $\mathcal{S}_2$ passing through $\boldsymbol{p}_2$. The function from the first tangent plane to the second is called the **differential** of F at $\boldsymbol{p}_1$ and is denoted $\mathrm{d}F$. The differential is defined by

$$\mathrm{d}F(\gamma_1'(t_0)) = \gamma_2'(t_0)$$

where $\gamma_2(t) = F(\gamma_1(t))$.

In this calculus situation, the chain rule tells us that $\mathrm{d}F$ is actually a linear transformation, and its matrix is

$$\begin{bmatrix} \frac{\partial f_1}{\partial x} & \frac{\partial f_1}{\partial y} & \frac{\partial f_1}{\partial z} \\ \frac{\partial f_2}{\partial x} & \frac{\partial f_2}{\partial y} & \frac{\partial f_2}{\partial z} \\ \frac{\partial f_3}{\partial x} & \frac{\partial f_3}{\partial y} & \frac{\partial f_3}{\partial z} \end{bmatrix},$$

where the partial derivatives are all evaluated at $\gamma_1(t_0) = \boldsymbol{p}_1$. This matrix is called the **Jacobian matrix** of F at $\boldsymbol{p}_1$.

Problem 3.3.2. (S)

a. Return to the special case of (3.3.1) and calculate the Jacobian matrix of F at $\boldsymbol{p}_1 = (3, 0, 3) = \gamma_1(t_0)$. Check that multiplying this matrix times the column vector $\gamma_1'(t_0)$ gives the column vector $\gamma_2'(t_0)$.

b. Now consider the more general case. Suppose $\gamma_1(t) = (a(t), b(t), c(t))$. Check that multiplying the general Jacobian matrix above by the column vector $\boldsymbol{p}_1 = \gamma_1'(t_0) = (a'(t_0), b'(t_0), c'(t_0))$ really does give $\gamma_2'(t_0)$. (Apply the appropriate form of the multivariable chain rule—remember the prime means the derivative with respect to t.)

The same idea generalizes to functions of any number of variables. Suppose N and M are any two positive integers and $F\colon \mathbb{R}^N \to \mathbb{R}^M$ is a smooth function. If $\gamma_1\colon \mathbb{R} \to \mathbb{R}^N$ is a smooth curve in $\mathbb{R}^N$ passing through $\boldsymbol{p}_1 = \gamma_1(t_0)$, then $\gamma_2(t) = F(\gamma_1(t))$ gives a smooth curve in $\mathbb{R}^M$ passing through $\boldsymbol{p}_2 = \gamma_2(t_0)$. We write

$$F(x_1, \ldots, x_N) = (f_1(x_1, \ldots, x_N), \ldots, f_M(x_1, \ldots, x_N)).$$

Since F is smooth, so are the f_j. The **differential** of F at $\boldsymbol{p}_1$ is the linear transformation $\mathrm{d}F\colon \mathbb{R}^N \to \mathbb{R}^M$ taking the tangent vector $\gamma_1'(t_0)$ to $\gamma_2'(t_0)$. By the chain rule, the matrix of $\mathrm{d}F$ (with respect to the standard bases) is

$$\begin{bmatrix} \frac{\partial f_1}{\partial x_1} & \cdots & \frac{\partial f_1}{\partial x_N} \\ \vdots & \vdots & \vdots \\ \frac{\partial f_M}{\partial x_1} & \cdots & \frac{\partial f_M}{\partial x_N} \end{bmatrix},$$

where all the partial derivatives are evaluated at $\boldsymbol{p}_1$. As in the special case we considered earlier with $N = M = 3$, this matrix is called the **Jacobian matrix** of F at $\boldsymbol{p}_1$.

Problem 3.3.3.

Let $F\colon \mathbb{R}^2 \to \mathbb{R}^2$ be defined by $F(x, y) = (f_1(x, y,), f_2(x, y)) = (e^{x+y}, e^{x-y})$.

a. Find the Jacobian matrix of F at $\boldsymbol{p}_1 = (0, 0)$.

b. Let $\gamma_1(t) = (\cos(t) - 1, \sin(t))$. Verify that $\gamma_1(0) = \boldsymbol{p}_1$ and find $\gamma_1'(0)$.

c. Let $\gamma_2(t) = F(\gamma_1(t))$. Find $\gamma_2'(0)$ using the Jacobian matrix of F at $\boldsymbol{p}_1$.

d. Check that the result of (c) agrees with what you would get by calculating $\gamma_2(t) = F(\gamma_1(t))$, then differentiating $\gamma_2(t)$ with respect to t, and finally evaluating $\gamma_2'(t)$ at $t = 0$.

Problem 3.3.4. (S)

Define $F\colon \mathbb{R}^2 \to \mathbb{R}^3$ by $F(x_1, x_2) = (x_1 \sin(x_2), \sin(x_1 x_2), x_1^2 + x_2)$.

a. Find the Jacobian matrix of F at $(0, \pi)$.

b. Use the matrix in (a) to calculate $\mathrm{d}F(1, 3)$.

The next problem is a reminder that in calculus we use differentials for *linear approximation*. For a differentiable function $f\colon \mathbb{R} \to \mathbb{R}$, the graph of the (linear) function $y = f(a) + f'(a)(x - a)$ is the line tangent to the graph of $y = f(x)$ at $x = a$, and for x near a, $f(x) \approx f(a) + f'(a)(x - a)$. There is a linear transformation $\mathbb{R}^1 \to \mathbb{R}^1$ lurking in this familiar fact. If we write $\Delta x = x - a$ and $\Delta y = f(x) - f(a)$, then the linear transformation $\mathrm{d}f$ is given by $\mathrm{d}f(\Delta x) = f'(a)\Delta x$, and the linear approximation is expressed by saying $\Delta y \approx \mathrm{d}f(\Delta x)$.

Problem 3.3.5. (S)
Consider the function $F\colon \mathbb{R}^3 \to \mathbb{R}^2$ defined by $F(x, y, z) = (x^2 \sin(y), xz^3)$.
a. Find the matrix of the differential $\mathrm{d}F$ at $(1, 0, 1)$.
b. Using the matrix from (a), what is $\mathrm{d}F(-0.1, 0.1, 0.1)$?
c. Compare $F(0.9, 0.1, 1.1) - F(1, 0, 1)$ and $\mathrm{d}F(-0.1, 0.1, 0.1)$.

It's time to return to Lie groups. Think of an n by n matrix as if it were an n^2-tuple by imagining all the entries written in one giant row (or column). So we think of a group of matrices as a set of points in $\mathbb{R}^N$, where $N = n^2$. To make the calculus analogy, think of our groups as if they were "surfaces." In our setting, the functions F of interest to us will be group homomorphisms. The next three problems give additional practice with the calculus definition of a differential in the matrix group context.

Problem 3.3.6.
Fix $n > 0$ and define $F\colon SL(n, \mathbb{R}) \to SL(n, \mathbb{R})$ by $F(X) = (X^{-1})^T$. We know from (3.2.6) that F is a group isomorphism. Restrict to the special case $n = 2$. Recall that for X in $SL(2, \mathbb{R})$,

$$X = \begin{bmatrix} x_{11} & x_{12} \\ x_{21} & x_{22} \end{bmatrix} \quad \text{implies} \quad X^{-1} = \begin{bmatrix} x_{22} & -x_{12} \\ -x_{21} & x_{11} \end{bmatrix}.$$

a. Write the matrix X as a 4-tuple $(x_{11}, x_{12}, x_{21}, x_{22})$, and think of $F\colon \mathbb{R}^4 \to \mathbb{R}^4$. Write $F(x_{11}, x_{12}, x_{21}, x_{22})$ as a 4-tuple. (You should find that F is a linear function, so it is certainly smooth.)
b. **(S)** Using the notation of (a), find the Jacobian matrix of F. Because F is linear, you should get a constant matrix, so it's not necessary to take the extra step of evaluating the partial derivatives at $I_2 = (1, 0, 0, 1)$. The result is the matrix of $\mathrm{d}F$ at $I_2 = (1, 0, 0, 1)$.
c. Let $A \in L(G)$ for $G = SL(2, \mathbb{R})$.

(i) Using the result of (3.1.11), write A as a 4-tuple.

(ii) Use the matrix of $\mathrm{d}F$ from (b) to find $\mathrm{d}F(A)$.

(iii) Verify that $\mathrm{d}F(A)$ is in $L(G)$.

d. Go back to matrix notation and write the 4-tuple $\mathrm{d}F(A)$ as a 2 by 2 matrix. Can you give a simple description of the matrix $\mathrm{d}F(A)$ in terms of the matrix A?

Problem 3.3.7.
This is another example of a homomorphism from $G = SL(2, \mathbb{R})$ to itself. Choose the matrix

$$M = \begin{bmatrix} 2 & 3 \\ 0 & 2 \end{bmatrix},$$

and define $F\colon SL(2, \mathbb{R}) \to SL(2, \mathbb{R})$ by $F(X) = MXM^{-1}$.
a. How do you know that if $X \in SL(2, \mathbb{R})$, then $F(X) \in SL(2, \mathbb{R})$?
b. Show that F is a group homomorphism.
c. Proceeding as in (3.3.6), find a matrix for $\mathrm{d}F$ at I_2.

d. Use your matrix in (c) to find $dF(A)$ for $A \in L(G)$. (This time it's not so easy to see what the matrix description of dF is; we'll return to this situation in Chapter 4.)

Fix positive integers n and m and assume $G_1 \subseteq GL(n, \mathbb{R})$ and $G_2 \subseteq GL(m, \mathbb{R})$ are Lie groups. Also assume $F: G_1 \to G_2$ is a group homomorphism. Let $N = n^2$ and $M = m^2$. If, when we think of F as if it were a function $\mathbb{R}^N \to \mathbb{R}^M$, F is smooth, we say that F is a **Lie group homomorphism**. In other words, Lie groups have two kinds of structure. They are groups, and a Lie group homomorphism "preserves" the group structure. But also, they are groups on which we can do calculus, and by being smooth, a Lie group homomorphism "preserves" the calculus structure as well. Thus invertible Lie group homomorphisms are *symmetries* of Lie groups. If F is a one-to-one onto Lie group homomorphism, we call it a **Lie group isomorphism** and say the two Lie groups are **isomorphic**—we might say isomorphic *as Lie groups* for emphasis. (As for group isomorphisms, the inverse of a Lie group isomorphism is again a Lie group isomorphism, so it doesn't matter which Lie group is the domain of the isomorphism.)

Problem 3.3.8.

Each part of this problem is a special case of what we will see later is a general theorem.

a. **(S)** As in (3.2.3), Let $F: GL(2, \mathbb{R}) \to G_2$ be given by $F(X) = \det(X)$; we know F is a group homomorphism. If $X = (x_{ij})$, then $\det(X) = x_{11}x_{22} - x_{12}x_{21}$. Using this formula, show that the differential dF at the identity I_2 takes the matrix $X \in \mathcal{M}(2, \mathbb{R})$ to $\operatorname{tr}(X) \in \mathbb{R}$, where $\operatorname{tr}(X) = x_{11} + x_{22}$. In calculating the differential, write the matrix X as a 4-tuple in the following order: $(x_{11}, x_{12}, x_{21}, x_{22})$.

b. Now go up a dimension and let $F: GL(3, \mathbb{R}) \to G_2$ be given by $F(X) = \det(X)$. As in (a), F is a group homomorphism by (3.2.3). If $X = (x_{ij})$, then $\det(X)$ is given by the following formula:

$$\det(X) = x_{11}x_{22}x_{33} + x_{12}x_{23}x_{31} + x_{13}x_{21}x_{32} - x_{13}x_{22}x_{31} - x_{12}x_{21}x_{33} - x_{11}x_{23}x_{32}.$$

Using this formula, show that the differential dF at the identity I_3 takes the matrix $X \in \mathcal{M}(3, \mathbb{R})$ to $\operatorname{tr}(X) \in \mathbb{R}$, where $\operatorname{tr}(X) = x_{11} + x_{22} + x_{33}$. In calculating the differential, write the matrix X as a 9-tuple in the following order:

$$(x_{11}, x_{12}, x_{13}, x_{21}, x_{22}, x_{23}, x_{31}, x_{32}, x_{33}).$$

Assume $F: G_1 \to G_2$ is a Lie group homomorphism. As before, we think of the n by n matrices in G_1 as points in $\mathbb{R}^N$ for $N = n^2$, and we think of the m by m matrices in G_2 as points in $\mathbb{R}^M$ for $M = m^2$. We know I_n is in G_1 and I_m is in G_2 (why?), and by (3.2.10), since F is a group homomorphism, we know $F(I_n) = I_m$. We will define the *differential of F at the identity* as a function from $L(G_1)$ to $L(G_2)$ following the same procedure as in our calculus examples. Recall that $L(G_1)$ is a subspace of $\mathcal{M}(n, \mathbb{R})$, which we think of as $\mathbb{R}^N$, and $L(G_2)$ is a subspace of $\mathcal{M}(m, \mathbb{R})$, which we think of as $\mathbb{R}^M$.

To make our definition in the group setting, we set up our notation so there is a subscript 1 on everything concerning the domain group G_1 and a 2 on everything concerning the range group G_2. Suppose A_1 is an arbitrary element of $L(G_1)$. By our definition of the tangent space $L(G_1)$, $A_1 \in L(G_1)$ means there is a smooth curve $\alpha_1 : \mathbb{R} \to G_1$ with $\alpha_1(0) = I_n$ and $\alpha_1'(0) = A_1$. Let $\alpha_2(t) = F(\alpha_1(t))$. Define

$$\mathrm{d}F(A_1) = \alpha_2'(0).$$

This is our formal definition of the **differential of F at the identity**.

Should we worry that the definition of $\mathrm{d}F(A_1)$ seems to depend on the choice of the curve α_1? What if some *other* curve β_1 also passes through I_n and satisfies $\beta_1'(0) = A_1$? If we let $\beta_2(t) = F(\beta_1(t))$ and calculate $\beta_2'(0)$, will we get the same result? Yes we *will*. The reason is the chain rule, which tells us that the derivative $\beta_2'(0)$ is determined by the derivative(s) of F and by $\beta_1'(0) = \alpha_1'(0)$, exactly the same ingredients determining $\alpha_2'(0)$.

In (3.3.9) below, you'll verify that this definition of $\mathrm{d}F$ really does give a function from $L(G_1)$ to $L(G_2)$, and in (3.3.10), you'll show that this function really is a linear transformation. (With no explicit calculus-style formula for F, we can't just appeal to a Jacobian matrix in making our argument.)

Problem 3.3.9.

Continue the notation above.

a. Explain why α_2 is a smooth curve in G_2 passing through the identity I_m.

b. Explain why $\mathrm{d}F(A_1) = \alpha_2'(0)$ is in $L(G_2)$.

Problem 3.3.10.

Continue the notation of (3.3.9). In this problem, you'll show that $\mathrm{d}F : L(G_1) \to L(G_2)$ is actually a linear transformation. You will find (3.1.6) and (3.1.8) useful. As there, you can assume that the product rule holds for the derivatives of matrix-valued functions.

a. Suppose A_1 and B_1 are in $L(G_1)$ with $A_1 = \alpha_1'(0)$ and $B_1 = \beta_1'(0)$, where α_1 and β_1 are smooth curves passing through the identity I_n in G_1. The goal of this part and the next is to show that $\mathrm{d}F(A_1 + B_1) = \mathrm{d}F(A_1) + \mathrm{d}F(B_1)$. To accomplish this, we'll need one more curve in G_1 and three curves in G_2:

$$\begin{aligned} \gamma_1(t) &= \alpha_1(t)\beta_1(t), \\ \alpha_2(t) &= F(\alpha_1(t)), \\ \beta_2(t) &= F(\beta_1(t)), \\ \gamma_2(t) &= F(\gamma_1(t)) = F(\alpha_1(t)\beta_1(t)). \end{aligned}$$

(By the way, the names of these Greek letters are alpha (α), beta (β), and gamma (γ).)

(i) Explain why the curve γ_1 is in G_1 and passes through the identity of G_1.

(ii) Explain why the curve γ_2 is in G_2 and passes through the identity of G_2.

(iii) Show $\gamma_2'(0) = \alpha_2'(0) + \beta_2'(0)$.

b. Explain why result (iii) of (a) tells you that $\mathrm{d}F(A_1 + B_1) = \mathrm{d}F(A_1) + \mathrm{d}F(B_1)$ for all $A_1, B_1 \in L(G_1)$.

c. Show that $\mathrm{d}F(cA_1) = c\,\mathrm{d}F(A_1)$ for every $c \in \mathbb{R}$ and every $A_1 \in L(G_1)$. (This time, assume $A_1 = \alpha_1'(0)$, $\alpha_2(t) = F(\alpha_1(t))$ and $\gamma_2(t) = F(\alpha_1(ct))$.)

An important example of a homomorphism is the determinant

$$F = \det : GL(n, \mathbb{R}) \to GL(1, \mathbb{R}),$$

(where we identify the group of nonzero real numbers under multiplication with the isomorphic group of invertible 1 by 1 matrices under matrix multiplication as in (3.2.2)). From (3.1.15), the tangent space at the identity for $GL(n, \mathbb{R})$ is $\mathcal{M}(n, \mathbb{R})$. The $n = 3$ case of (3.3.8) is actually true in general and is an important property of the determinant function: the differential of the determinant is the trace

$$\mathrm{d}F = \mathrm{tr} : \mathcal{M}(n, \mathbb{R}) \to \mathcal{M}(1, \mathbb{R}),$$

(where we identify the isomorphic one-dimensional vector spaces $\mathbb{R}$ and $\mathcal{M}(1, \mathbb{R})$). In other words, if $F(M) = \det(M)$ for $M \in GL(n, \mathbb{R})$, then $\mathrm{d}F(A) = \mathrm{tr}(A)$ for $A \in \mathcal{M}(n, \mathbb{R})$.

In order to see why this property holds in general, we need the definition of the determinant of an n by n matrix for an arbitrary positive integer n. We can use the example of the determinant of the 3 by 3 matrix X to explain the definition in the general case. If $X = (x_{ij})$ for $i, j = 1, 2, 3$, then

$$\det(X) = x_{11}x_{22}x_{33} + x_{12}x_{23}x_{31} + x_{13}x_{21}x_{32} - x_{13}x_{22}x_{31} - x_{12}x_{21}x_{33} - x_{11}x_{23}x_{32}.$$

Notice the pattern in the formula for $\det(X)$. Each term in the sum has three factors, one from each row and one from each column. The factors are written so that their first subscripts are in the order 1,2,3. The three second subscripts in each term give some permutation of 1,2,3. The three terms with plus signs correspond to the permutations 1,2,3; 2,3,1; 3,1,2. The three terms with minus signs correspond to the permutations 3,2,1; 2,1,3; 1,3,2. The first three permutations are called *even* permutations, and the second three are called *odd.* For example, the second of these positive terms corresponds to the permutation $1 \mapsto 2$, $2 \mapsto 3$, $3 \mapsto 1$. This permutation is called *even* because it can be achieved by first interchanging 1 and 3 and then interchanging 2 and 3—*two* interchanges, an *even* number of interchanges. If π represents this permutation, we can write $\pi(1) = 2$, $\pi(2) = 3$, $\pi(3) = 1$, so the second positive term can be written $x_{1\pi(1)}x_{2\pi(2)}x_{3\pi(3)}$. The permutation 1,2,3 is also even, because it can be achieved by zero interchanges, and zero is an even number. (It can also be achieved by 2 interchanges: interchange 1 and 2 and then interchange them again.) Similarly, the first permutation with a minus sign is $1 \mapsto 3$, $2 \mapsto 2$, $3 \mapsto 1$. This permutation is called *odd* because it can be accomplished by *one* interchange (interchange 1 and 3), and one is an *odd* number.

Remark You should be worried by the fact, noted above, that a particular permutation is expressible as a composition of interchanges in more than one way. What does that ambiguity do to the assignment of the plus or minus sign in our formula? Fortunately, there is a theorem that assures that *every* decomposition of a permutation into interchanges has the same parity—it is not possible to write the same permutation as a composition of an even number of interchanges and also as a composition of an odd number of interchanges.

With just a bit more notation, we can write the definition of the determinant of an arbitrary n by n matrix $X = (x_{ij})$. Let $\mathcal{S}_n$ be the set of *all* permutations of 1,2, $\ldots, n$.

(S_n contains $n!$ permutations, where $n! = n(n-1)(n-2)\cdots 2\cdot 1$.) Then

$$\det(X) = \sum_{\pi\in S_n} (\pm 1)x_{1\pi(1)}x_{2\pi(2)}\cdots x_{n\pi(n)}.$$

When π is an even permutation the sign is positive and when π is odd the sign is negative.

Problem 3.3.11.
Check that the general definition above gives the familiar result for $n = 2$.

Problem 3.3.12. (C)
Use the general formula above for $\det(X)$ to show that the trace is the differential of the determinant. That is, let

$$F = \det : GL(n, \mathbb{R}) \to GL(1, \mathbb{R}),$$

and let dF be the differential of F at the identity I_n. Show that $dF(A) = \mathrm{tr}(A)$ for all A in $\mathcal{M}(n, \mathbb{R})$.

Problem 3.3.13.
Use (3.3.12) to show that if $\gamma(t) = (a_{ij}(t))$ is a smooth curve in $SL(n, \mathbb{R})$ passing through I_n, then $\mathrm{tr}(\gamma'(0)) = 0$.

The next problem shows that it doesn't matter whether you compose functions first and then calculate the differential or calculate differentials first and then compose.

Problem 3.3.14.
Suppose G_1, G_2 and G_3 are Lie (matrix) groups and

$$F_1 : G_1 \to G_2,$$
$$F_2 : G_2 \to G_3$$

are Lie group homomorphisms. We know that dF_1 and dF_2 are linear transformations,

$$dF_1 : L(G_1) \to L(G_2),$$
$$dF_2 : L(G_2) \to L(G_3).$$

a. Explain why $dF_2 \circ dF_1$ is a linear transformation from $L(G_1)$ to $L(G_3)$.
b. Explain why $d(F_2 \circ F_1)$ is a linear transformation from $L(G_1)$ to $L(G_3)$.
c. Show that

$$d(F_2 \circ F_1) = dF_2 \circ dF_1.$$

(Hint: To show that two functions are equal, you have to apply each of them to a general element of their common domain, and then show that they each give the same output. In this case, choose a general element $A_1 = \alpha_1'(0)$ in $L(G_1)$.)

Problem 3.3.15.
This problem shows that isomorphic Lie groups have isomorphic tangent spaces. You can assume that if a Lie group homomorphism is invertible, then its inverse is again a Lie group homomorphism. Show that if G_1 and G_2 are matrix groups and F is an invertible

Lie group homomorphism

$$F: G_1 \to G_2,$$

then the differential

$$\mathrm{d}F: L(G_1) \to L(G_2)$$

is also invertible.

3.4 Putting the pieces together

3.4.1 Tangent spaces

Fix a positive integer n. Suppose G is a group of n by n matrices and $\gamma: \mathbb{R} \to G$ is a function. We can write $\gamma(t) = (\gamma_{ij}(t))$, where $\gamma_{ij}(t)$ is the i, j entry of the matrix $\gamma(t)$. The coordinate functions γ_{ij} are familiar kinds of functions, $\gamma_{ij}: \mathbb{R} \to \mathbb{R}$. We say the function γ is **smooth** if the functions γ_{ij} are smooth, that is, if they are infinitely differentiable. Then the **tangent vector** to G at $\gamma(0)$ defined by γ is the n by n matrix $\gamma'(0) = (\gamma'_{ij}(0))$. Borrowing the language of 2 or 3 dimensions, we call γ a *smooth curve* in G, and we say the curve *passes through* $\gamma(0)$.

The **tangent space at the identity** $L(G)$ for the matrix group G consists of all tangent vectors $\gamma'(0)$, where γ is a smooth function satisfying $\gamma: \mathbb{R} \to G$ and $\gamma(0) = I_n$,

$$L(G) = \{\gamma'(0) \, : \, \gamma : \mathbb{R} \to G, \gamma(0) = I_n, \gamma \text{ smooth } \}.$$

Theorem 3.1 *Fix $n > 0$. For G a group of n by n matrices, the tangent space $L(G)$ is a subspace of the vector space $\mathcal{M}(n, \mathbb{R})$ of all n by n matrices.*

Remark Although both G and $L(G)$ consist of n by n matrices, G is a *group*, and the group operation in G is *matrix multiplication*, while $L(G)$ is a *vector space* over $\mathbb{R}$, and the vector space operations in $L(G)$ are *matrix addition* and multiplication of a matrix by a scalar from $\mathbb{R}$.

Theorem 3.2 *Fix $n > 0$. We have the following special cases. (Some of these results won't be fully established until Chapter 4.)*

1. *If $G = GL(n, \mathbb{R})$, then $L(G) = \mathcal{M}(n, \mathbb{R})$. The dimension of $L(G)$ is n^2.*
2. *If $G = SL(n, \mathbb{R})$, then $L(G)$ consists of those matrices in $\mathcal{M}(n, \mathbb{R})$ which have trace equal to zero. The dimension of $L(G)$ is $n^2 - 1$.*
3. *If G is $O(2, \mathbb{R})$ in the Euclidean case, then*

   $$L(G) = \left\{ \begin{bmatrix} 0 & b \\ -b & 0 \end{bmatrix} : b \in \mathbb{R} \right\}.$$

   The dimension of $L(G)$ is 1.
4. *If $G = \mathcal{L}(2, \mathbb{R})$, then*

   $$L(G) = \left\{ \begin{bmatrix} 0 & b \\ b & 0 \end{bmatrix} : b \in \mathbb{R} \right\}.$$

   The dimension of $L(G)$ is 1.

5. *If $G = O(n, \mathbb{R})$ in the Euclidean case, then $L(G)$ consists of the n by n* ***skew-symmetric*** *matrices*

$$L(G) = \{B \in \mathcal{M}(n, \mathbb{R}) \ : \ B^T = -B\}.$$

The dimension of $L(G)$ is $(n^2 - n)/2$. (Note: The dimension has to be an integer. Why is $n^2 - n$ always even?)

6. *If $G = SO(n, \mathbb{R})$ in the Euclidean case, then $L(G)$ consists of the n by n skew-symmetric matrices — i.e., $O(n, \mathbb{R})$ and $SO(n, \mathbb{R})$ have the same tangent space at the identity.*

3.4.2 Group homomorphisms

Invertible group homomorphisms are *symmetries* of groups. Suppose we have a group G_1 with binary operation $*$ and a group G_2 with binary operation $\diamond$. A function $F: G_1 \to G_2$ satisfying

$$F(g * h) = F(g) \diamond F(h)$$

for all $g, h \in G_1$ is a **group homomorphism**. If the function F is also one-to-one and onto, we say it is a **group isomorphism**, and we say that groups G_1 and G_2 are **isomorphic**.

Proposition 3.3 *If G_1 and G_2 are groups and $F: G_1 \to G_2$ is a group homomorphism, then F takes the identity of G_1 to the identity of G_2. In the special case when G_1 and G_2 are matrix groups, of n by n and m by m matrices respectively, then $F(I_n) = I_m$.*

Proposition 3.4 *The inverse of a group isomorphism is again a group isomorphism.*

Proposition 3.5

a. *The groups S^3 and $SU(2, \mathbb{C})$ are isomorphic.*

b. *There is a group homomorphism from S^3 onto $SO(3, \mathbb{R})$, but it is not one-to-one. (It is two-to-one.)*

c. *There is a group homomorphism from $\mathbb{R}$ onto S^1, but it is not one-to-one. (Infinitely many real numbers map to each complex number in S^1.)*

3.4.3 Differentials

Fix positive integers n and m and let $N = n^2$ and $M = m^2$. Suppose $G_1 \subseteq GL(n, \mathbb{R})$ and $G_2 \subseteq GL(m, \mathbb{R})$ are Lie groups and $F: G_1 \to G_2$ is a group homomorphism and is also smooth (regarded as a map from $\mathbb{R}^N$ to $\mathbb{R}^M$). Then we say F is a **Lie group homomorphism**. A one-to-one and onto Lie group homomorphism is called a **Lie group isomorphism**, and we say the Lie groups G_1 and G_2 are **isomorphic**; for emphasis, we may say **isomorphic as Lie groups**. A Lie group isomorphism $G \to G$ is a *symmetry* of the Lie group G, since it "preserves" both the group operation and also the calculus structure of G.

Suppose G_1 and G_2 are Lie groups, with tangent spaces at the identity $L(G_1)$ and $L(G_2)$ respectively. The differential provides a relationship between Lie group homomorphisms from G_1 to G_2 and linear transformations from $L(G_1)$ to $L(G_2)$. These ideas lead us to some of the great theorems of the Lie theory, discussed in the next section.

Suppose G_1 and G_2 are Lie groups and $F: G_1 \to G_2$ is a Lie group homomorphism. The **differential** of F **at the identity** is a function

$$\mathrm{d}F: L(G_1) \to L(G_2)$$

defined as follows: If $A = \alpha_1'(0)$ is a tangent vector in $L(G_1)$ for a smooth curve α_1 passing through the identity in G_1, then $\mathrm{d}F$ takes A to

$$\mathrm{d}F(A) = \alpha_2'(0) \in L(G_2),$$

where $\alpha_2(t) = F(\alpha_1(t))$ for all t in $\mathbb{R}$.

Theorem 3.6 *Suppose G_1 and G_2 are two Lie groups with tangent spaces at the identity $L(G_1)$ and $L(G_2)$ respectively. Suppose*

$$F: G_1 \to G_2$$

is a Lie group homomorphism. Then

$$dF: L(G_1) \to L(G_2)$$

is a linear transformation.

Theorem 3.7 *Suppose G_1, G_2 and G_3 are Lie groups and $F_1: G_1 \to G_2$ and $F_2: G_2 \to G_3$ are Lie group homomorphisms. Then*

$$d(F_2 \circ F_1) = dF_2 \circ dF_1.$$

In other words, it doesn't matter whether you compose functions first and then find the differential of the composite or find differentials first and then compose them.

If you did (3.3.14), then you may have noticed that the crux of the proof of Theorem 3.7 is the fact that

$$(F_2 \circ F_1) \circ \gamma_1 = F_2 \circ (F_1 \circ \gamma_1),$$

where $\gamma_1: \mathbb{R} \to G_1$ is a curve passing through the identity in G_1. In other words, the key fact is that composition of functions is associative.

Corollary 3.8 *If G_1 and G_2 are Lie groups and F is an invertible Lie group homomorphism*

$$F: G_1 \to G_2$$

then the differential

$$dF: L(G_1) \to L(G_2)$$

is also invertible. In other words,

Isomorphic Lie groups have isomorphic tangent spaces.

In particular, isomorphic matrix groups have tangent spaces of the same dimension. So, the dimension of the tangent space captures some important information about the structure of the group. The ***dimension*** *of a matrix group G is the dimension of its tangent space at the identity:*

$$\dim G = \dim L(G).$$

Of course, we know that the dimension doesn't capture *all* the information about the group. Not only can a subgroup have the same tangent space as the containing group (think of the Euclidean orthogonal group and its subgroup of rotations), but also quite different matrix groups can have tangent spaces of the same dimension (think of the Euclidean orthogonal group $O(3, \mathbb{R})$ and $SL(2, \mathbb{R})$). So we have more work to do on tangent spaces, and we will do that work in Chapter 5.

Suggestions for further reading

There are many places to read more about group theory. One is Artin, M., *Algebra*, Prentice Hall (1991). See chapter 2 (Groups), chapter 5 (Symmetry), and chapter 6 (More Group Theory).

Singer, S.F., *Linearity, Symmetry, and Prediction in the Hydrogen Atom*, Undergraduate Texts in Mathematics, Springer (2005). Students who already know some group theory might enjoy chapter 4 on Lie groups and Lie group representations.

3.5 A broader view: Hilbert's fifth problem

The differential of a Lie group homomorphism captures important information about the homomorphism itself. We can't prove it using the methods of this text—it depends on a property from topology called being "simply connected," but the following theorem is true. It is one of the great theorems of the Lie theory.

Theorem 3.9* *Assume G_1 is a simply connected Lie group, and suppose F and H are Lie group homomorphisms from G_1 to another Lie group G_2. Assume further that $dF = dH$. Then $F = H$.*

In other words, provided the domain Lie group is "nice," if two Lie group homomorphisms have the same differential, they are the same homomorphism.

So what does "nice" mean here? To answer it, we must dip into topology. (For a quick and accessible overview of these topological ideas, see pages 173–203 in *Naive Lie Theory* by John Stillwell.) A group is *path-connected* if any two group elements can be connected by a continuous curve in the group. The groups $SL(n, \mathbb{R})$, $SO(n, \mathbb{R})$ and $\mathcal{L}^{++}(2, \mathbb{R})$ are path-connected. However, the groups $GL(n, \mathbb{R})$, $O(n, \mathbb{R})$ and $\mathcal{L}(2, \mathbb{R})$ are *not* path-connected, because there is no way to continuously connect a matrix of determinant 1 to a matrix of determinant -1 and stay within the larger group.

Roughly speaking, a group is simply connected if it has no "holes." More formally, a group is *simply connected* if it is path-connected and any closed continuous curve in the group can be continuously shrunk to a point. The real line $\mathbb{R}$ and the entire plane $\mathbb{R}^2$ (each a group under addition) are simply connected. However, the multiplicative group S^1 of complex numbers of length 1 (the unit circle in the plane) is connected but not simply connected. On the other hand, the multiplicative subgroup of quaternions of length 1

$$S^3 = \{a + bi + cj + dk \, : \, a, b, c, d \in \mathbb{R}, a^2 + b^2 + c^2 + d^2 = 1\}$$

is simply connected, as are the matrix groups $SU(n, \mathbb{C})$ for $n \geq 2$. The matrix groups $SO(n, \mathbb{R})$ and $SL(n, \mathbb{R})$ are not simply connected for $n > 1$, although both are path-connected. You can see the "hole" in $SO(2, \mathbb{R})$ when its elements are identified with the points of S^1, as in section 2.1. Similarly, you can "see" that $SU(2, \mathbb{C})$ is simply connected via the isomorphism between $SU(2, \mathbb{C})$ and S^3 in (3.2.13).

There is another of the great theorems of the subject that should be mentioned in this context. The definition of a Lie group homomorphism is actually stronger than it needs to be.

Theorem 3.10* *If G_1 and G_2 are Lie groups and $F : G_1 \to G_2$ is a continuous group homomorphism, then F is infinitely differentiable.*

This is certainly *not* true for ordinary calculus functions. (What is an example of a continuous function $\mathbb{R} \to \mathbb{R}$ which is not differentiable?) It is the assumption that F is a group homomorphism that, together with continuity, gives enough to force differentiability. We will look at some special cases of this result in Chapter 4.

Theorem 3.10* is closely related to one of Hilbert's famous "twenty three problems." David Hilbert (1862-1943) was one of the — perhaps *the* — pre-eminent figures of late nineteenth century/early twentieth century mathematics. In his address to the International Congress of Mathematicians, held in Paris in 1900, Hilbert described twenty three problems that he thought should be the goals of mathematics in the twentieth century. And, to a very large extent, they were.

Hilbert's fifth problem is usually stated as follows:

Under what conditions is a *topological group* also a *Lie group*?

Loosely speaking, a topological group is one where there is a notion of "closeness," so that it is possible to define a continuous function from the group to itself. (Informally, a continuous function is one that can produce output values that are arbitrarily close together when input values are chosen sufficiently close together.) A Lie group, on the other hand, is a group (like a matrix group) on which one can "do calculus." The fifth problem was solved in two papers in 1952 and 1953, the first by Gleason, Montgomery and Zippin, and the second by Yamabe.

It is impossible to resist including the following story about Hilbert and the distinguished mathematician Emmy Noether, drawn from Constance Reid's magisterial 1970 biography *Hilbert*. During World War I, Hilbert (along with his colleague Felix Klein) invited Emmy Noether to join them at the University of Göttingen because

> she had an impressive knowledge of certain subjects which Hilbert and Klein needed for their work on relativity theory, and they were both determined that she must stay in Göttingen. But in spite of the fact that Göttingen had been the first university in Germany to grant a doctoral degree to a woman, it was still not an easy matter to obtain *habilitation* for one [i.e., an appointment to a position as Privatdozent,

the very bottom rung of the academic ladder, via submitting a thesis beyond the doctoral dissertation]. The entire Philosophical Faculty, which included philosophers, philologists and historians as well as natural scientists and mathematicians, had to vote on the acceptance of the habilitation thesis. Particular opposition came from the non-mathematical members of the Faculty.

They argued formally: "How can it be allowed that a woman become a Privatdozent? Having become a Privatdozent, she can then become a professor and a member of the University Senate. Is it permitted that a woman enter the Senate?" They argued informally, "What will our soldiers think when they return to the University and find that they are expected to learn at the feet of a woman?"

Hilbert's legendary rejoinder was, "Meine Herren, I do not see that the sex of the candidate is an argument against her admission as a Privatdozent. After all, the Senate is not a bathhouse." Hilbert did not win the argument, but he contrived to keep Noether at Göttingen by sponsoring her lectures under his own name. She did eventually (1919) win appointment as Privatdozent, and, Reid writes, "One of the most fertile circles of research in post-war Göttingen revolved around Emmy Noether." Noether, who was Jewish, remained at Göttingen until forced to flee Germany in 1933. Her US safe haven was at Bryn Mawr College where, tragically, in 1935 she died after surgery.

CHAPTER 4

One-parameter subgroups and the exponential map

Our definition of the tangent space of a Lie group G relies on producing smooth curves passing through the identity by finding appropriate functions $\gamma\colon \mathbb{R} \to G$. So far, we don't have a general procedure for creating such curves. Remarkably, it turns out that by requiring our curves to be more special, we can achieve the generality we seek.

4.1 One-parameter subgroups

Since $\mathbb{R}$ is a group under addition, we can require $\gamma\colon \mathbb{R} \to G \subseteq GL(n, \mathbb{R})$ to be a group homomorphism. We will find that by requiring that γ be a group homomorphism, we can somewhat relax our requirement of infinite differentiability.

A **one-parameter subgroup** of a matrix group G is (the image of) a *continuous* group homomorphism $\gamma\colon \mathbb{R} \to G$. A bit oddly, perhaps, the function γ itself is called a one-parameter subgroup too.

Problem 4.1.1.
Verify that the following functions $\gamma\colon \mathbb{R} \to GL(2, \mathbb{R})$ are actually group homomorphisms.

a. $\gamma(t) = \begin{bmatrix} \cos(t) & -\sin(t) \\ \sin(t) & \cos(t) \end{bmatrix}$,

b. $\gamma(t) = \begin{bmatrix} \cosh(t) & \sinh(t) \\ \sinh(t) & \cosh(t) \end{bmatrix}$,

c. **(S)** $\gamma(t) = \begin{bmatrix} e^t & 0 \\ 0 & e^{-t} \end{bmatrix}$,

d. $\gamma(t) = \begin{bmatrix} 1 & t \\ 0 & 1 \end{bmatrix}$.

Problem 4.1.2.
Show that γ is a one-parameter subgroup of $SL(3, \mathbb{R})$, where

$$\gamma(t) = \begin{bmatrix} e^t & 0 & 0 \\ 0 & e^t & 0 \\ 0 & 0 & e^{-2t} \end{bmatrix}.$$

Problem 4.1.3.

Show that γ is a one-parameter subgroup of $O(3, \mathbb{R})$, where

$$\gamma(t) = \begin{bmatrix} \cos t & -\sin t & 0 \\ \sin t & \cos t & 0 \\ 0 & 0 & 1 \end{bmatrix}.$$

Problem 4.1.4

Fix a positive integer n. Explain why every one-parameter subgroup $\gamma: \mathbb{R} \to G \subseteq GL(n, \mathbb{R})$ satisfies $\gamma(0) = I_n$.

As noted, in the following sections we will show that our definition of a one-parameter subgroup γ is equivalent to an apparently stronger one requiring that γ be smooth. The means to that end is the exponential map. Indeed, it will turn out that the exponential map is the source of *all* the curves we'll need in order to determine tangent spaces.

4.2 The exponential map in dimension 1

Assume that γ is a one-parameter subgroup,

$$\gamma: \mathbb{R} \to GL(1, \mathbb{R}).$$

We've already seen that $GL(1, \mathbb{R})$ is isomorphic to the multiplicative group of nonzero real numbers. Under this isomorphism, the 1 by 1 matrix $[a]$ corresponds to the real number a, so the identity matrix corresponds to the number 1. We will think of $\gamma: \mathbb{R} \to \mathbb{R}$. Since $\det([a]) = a \neq 0$ for $[a] \in GL(1, \mathbb{R})$, we have that $\gamma(t) \neq 0$ for all t. The following problems will show that this assumption implies that for all $t \in \mathbb{R}$, $\gamma(t) = b^t$, where $b = \gamma(1)$. In particular, γ is a smooth function. So, even though we seem to start out with a "generic" function γ meeting our conditions, it turns out to be a function of a very specific kind.

Now we embark on the argument to show that $\gamma(t)$ is actually equal to b^t for $b = \gamma(1)$. Since we know $\gamma(t) \neq 0$ for all t, we know $b \neq 0$. We'd like to know $b > 0$, and this follows from the Intermediate Value Theorem.

Intermediate Value Theorem *Assume $f: \mathbb{R} \to \mathbb{R}$ is a continuous function, and suppose $t_1, t_2 \in \mathbb{R}$ with $f(t_1) = s_1 < f(t_2) = s_2$. If $s_1 < c < s_2$, then there is some choice of t between t_1 and t_2 for which $f(t) = c$.*

Problem 4.2.1. (S)

Assume $\gamma: \mathbb{R} \to \mathbb{R}$, γ is a continuous function, $\gamma(t) \neq 0$ for all t, and $\gamma(0) = 1$. Use the Intermediate Value Theorem to show that $\gamma(1) = b > 0$. (Hint: Explain why $b < 0$ would lead to a contradiction.)

The fact that $b > 0$ is important, because it implies that we can meaningfully calculate $b^{1/k}$ for any positive integer k, as in the following problem.

Problem 4.2.2.

Assume $\gamma\colon \mathbb{R} \to GL(1, \mathbb{R})$ is a group homomorphism and $\gamma(1) = b > 0$.

a. **(S)** Assume $a \in \mathbb{R}$. Use induction to show $\gamma(ma) = (\gamma(a))^m$ for all positive integers m.

b. Explain why $\gamma(m) = b^m$ for all positive integers m.

c. **(S)** Use $\gamma(m + (-m)) = \gamma(0) = 1$ to show $\gamma(-m) = b^{-m}$ for all negative integers $-m$.

d. Use $1 = k(1/k)$ to show $\gamma(1/k) = b^{1/k}$ for all positive integers k.

e. Explain why it now follows that $\gamma(m/k) = b^{m/k}$ for every rational number m/k. (One approach is to begin by showing that this is true for *positive* rational numbers.)

From (4.2.2), we know that $\gamma(t) = b^t$ is true for every rational number t. We need a further argument to get this conclusion when t is *any* real number.

Suppose t is an arbitrary real number. Every real number is the limit of a sequence of rational numbers (think of its decimal expansion, for example), so assume

$$t = \lim_{n\to\infty} t_n,$$

where the t_n are all rational numbers. Because γ is a continuous function, another result from calculus tells us that

$$\gamma(t) = \gamma\left(\lim_{n\to\infty} t_n\right) = \lim_{n\to\infty} \gamma(t_n).$$

We know that $\gamma(t_n) = b^{t_n}$, since t_n is a rational number. So we have

$$\gamma(t) = \lim_{n\to\infty} b^{t_n}.$$

But the exponential function b^t is known to be continuous, so by the reasoning we used for γ,

$$\gamma(t) = \lim_{n\to\infty} b^{t_n} = b^{\lim_{n\to\infty} t_n} = b^t.$$

Finally, we can conclude that $\gamma(t) = b^t$ for every real number t. □

It's easier to do calculus with exponential functions written using the base e, $\exp(t) = e^t$, so let $a = \ln(b)$. Now we can write $\gamma(t) = b^t = (\exp(\ln(b))^t = (\exp(a))^t = \exp(at)$, and $\gamma'(0) = a$ by the chain rule. In other words, we have proved the following fact.

> If $\gamma\colon \mathbb{R} \to GL(1, \mathbb{R})$ is a continuous group homomorphism, then $\gamma(t) = \exp(at)$ for $a = \gamma'(0)$.

Our focus has been on showing that if γ is both continuous and a group homomorphism, then γ is differentiable (in fact, smooth). The exponential function was just a means to that end. But it was a long road. We can obtain a shorter proof of the fact that γ is an exponential function if we strengthen the hypothesis and assume that γ is *differentiable* and a group homomorphism. Again, the proof requires a powerful theorem from calculus. The first step is the following problem.

Problem 4.2.3.
Assume γ is a differentiable group homomorphism,

$$\gamma\colon \mathbb{R} \to GL(1, \mathbb{R}).$$

Calculate $\gamma'(t)$ from the definition of the derivative as the limit of a difference quotient

$$\gamma'(t) = \lim_{h \to 0} \frac{1}{h}[\gamma(t+h) - \gamma(t)],$$

and show that $\gamma'(t) = \gamma'(0)\gamma(t)$.

In other words, (4.2.3) shows that $y = \gamma(t)$ is a solution of the initial value problem

$$\frac{dy}{dt} = ay \quad \text{and} \quad y(0) = 1,$$

where $a = \gamma'(0)$. From calculus, we know that $y = \exp(at)$ is also a solution. The powerful result from calculus that we need now is a theorem on *uniqueness* of solutions to an initial value problem like this one. Invoking this theorem, we can conclude that $\gamma(t) = \exp(at)$.

This tells us what we need to know for the very special case of 1 by 1 matrices. But we need to extend our results to n by n matrices for any choice of the postive integer n. It will turn out that an analogous result is true:

> If $\gamma\colon \mathbb{R} \to GL(n, \mathbb{R})$ is a continuous group homomorphism, then $\gamma(t) = \exp(tA)$ for $A = \gamma'(0)$.

It makes sense to multiply a matrix A by a scalar t, but what does $\exp(tA)$ mean?

4.3 Calculating the matrix exponential

Imitating the Taylor series for e^x, we define the exponential function for a matrix $A \in \mathcal{M}(n, \mathbb{R})$, where n is any positive integer, by a series

$$\exp(A) = I_n + A + \frac{1}{2!}A^2 + \cdots + \frac{1}{m!}A^m + \cdots.$$

In this section you'll calculate examples of matrix exponentials mechanically, using this definition without worrying about whether the series above converges. (It does, always!) For each of the problems below, just write out enough terms of the series that you can see the emerging pattern in the matrix entries. In some cases, you'll need to be able to recognize the Taylor series for some standard calculus functions (such as e^t, $\sin(t)$, $\cos(t)$, $\sinh(t)$, and $\cosh(t)$) in order to describe your results nicely.

Later, we'll need to know the fact that $\det(\exp(tA)) = e^{\mathrm{tr}(tA)}$. Recall that the *trace* $\mathrm{tr}(B)$ of a matrix B is the sum of its diagonal entries. It's surprising that the determinant of the exponential — which depends on *all* the entries of A — is completely determined by just the diagonal entries of A. That's why the following problems also ask you to calculate $\det(\exp(tA))$ and $e^{\mathrm{tr}(tA)}$ in these special cases.

Problem 4.3.1.
Let

$$A = \begin{bmatrix} 0 & 1 \\ 0 & 0 \end{bmatrix}.$$

a. Find $tA, (tA)^2, (tA)^3, (tA)^4, \ldots$.
b. Find $\exp(tA)$.
c. Check that $\gamma(t) = \exp(tA)$ implies $\gamma'(0) = A$ in this case.
d. Check that $\det(\exp(tA)) = e^{\mathrm{tr}(tA)}$ in this case.

Problem 4.3.2.
Let

$$A = \begin{bmatrix} 1 & 0 \\ 0 & -1 \end{bmatrix}.$$

a. Find $tA, (tA)^2, (tA)^3, (tA)^4, \ldots$.
b. **(S)** Find $\exp(tA)$.
c. Check that $\gamma(t) = \exp(tA)$ implies $\gamma'(0) = A$ in this case.
d. Check that $\det(\exp(tA)) = e^{\mathrm{tr}(tA)}$ in this case.

Problem 4.3.3.
Let

$$A = \begin{bmatrix} 0 & -1 \\ 1 & 0 \end{bmatrix}.$$

a. Find tA, $(tA)^2$, $(tA)^3, (tA)^4, \ldots$.
b. **(S)** Find $\exp(tA)$.
c. Check that $\gamma(t) = \exp(tA)$ implies $\gamma'(0) = A$ in this case.
d. Check that $\det(\exp(tA)) = e^{\mathrm{tr}(tA)}$ in this case.

Problem 4.3.4.
Let

$$A = \begin{bmatrix} 0 & 1 \\ 1 & 0 \end{bmatrix}.$$

a. Find tA, $(tA)^2$, $(tA)^3, (tA)^4, \ldots$.
b. Find $\exp(tA)$.
c. Check that $\gamma(t) = \exp(tA)$ implies $\gamma'(0) = A$ in this case.
d. Check that $\det(\exp(tA)) = e^{\mathrm{tr}(tA)}$ in this case.

Problem 4.3.5.
Show by an example that $\exp(A) = I$ needn't imply that A is the zero matrix. (Hint: Look at (4.3.3).)

Use a computer algebra system to do the remaining problems in this section. As an example, some useful commands in Maple appear at the end of this section.

Problem 4.3.6.

As a warmup, recalculate $\exp(tA)$ for

$$A = \begin{bmatrix} 0 & -1 \\ 1 & 0 \end{bmatrix}.$$

Problem 4.3.7.

Now work with a different matrix A. (In most computer algebra systems, when you change the value of the variable A, the values of the variables that depend on A are automatically modified.) Find $\exp(tA)$ for

$$A = \begin{bmatrix} 0 & 0 & 1 \\ 0 & 0 & 0 \\ -1 & 0 & 0 \end{bmatrix}.$$

Problem 4.3.8.

a. Find $M = \exp(tA)$ for

$$A = \begin{bmatrix} 0 & 1 \\ 1 & 0 \end{bmatrix}.$$

b. Check that $\exp(tA)$ is in $\mathcal{L}(2, \mathbb{R})$ by calculating $M^T C M$ for

$$C = \begin{bmatrix} 1 & 0 \\ 0 & -1 \end{bmatrix}.$$

Problem 4.3.9.

a. Find $\exp(tA)$ for

$$A = \begin{bmatrix} 0 & -1 & 0 \\ 1 & 0 & 0 \\ 0 & 0 & 0 \end{bmatrix}.$$

b. Check that $\exp(tA)$ is in the Euclidean orthogonal group.

Problem 4.3.10.

a. Find $\exp(tA)$ for

$$A = \begin{bmatrix} 0 & 1 & 0 & 0 \\ 1 & 0 & 0 & 0 \\ 0 & 0 & 0 & 0 \\ 0 & 0 & 0 & 0 \end{bmatrix}.$$

b. Check that $\exp(tA)$ is in $\mathcal{L}(4, \mathbb{R})$, using

$$C = \begin{bmatrix} 1 & 0 & 0 & 0 \\ 0 & -1 & 0 & 0 \\ 0 & 0 & -1 & 0 \\ 0 & 0 & 0 & -1 \end{bmatrix}.$$

Calculating the matrix exponential using Maple

These instructions are for Maple, but other software packages can also do these calculations with ease.

You can begin on a Worksheet, on a Classic Worksheet, or by choosing the Maple prompt (which is represented by [>) from the menu bar at the top of the page. When you open a Worksheet or choose the prompt icon, you will see the Maple prompt >.

Every line of instructions in Maple for which you want to see output should end with a semicolon. You can end a line with a colon to suppress the output. Use the symbol := to assign value to a variable. Note that Maple is case sensitive, so be exact in typing commands and be consistent in the use of upper and lower case letters in variable names.

The following sample walks you through the calculation in (3.3.6) of $\exp(tA)$ for

$$A = \begin{bmatrix} 0 & -1 \\ 1 & 0 \end{bmatrix},$$

using Maple—with some digressions to show you some other useful Maple commands.

In what follows, type exactly what is given after the Maple prompt. Comments in square brackets are for you, the human reader, not for the computer. (This is *not* the format for "comments" in Maple—don't type them into the computer.)

```
> with(linalg);
```

[This loads the special linear algebra package. You'll get a long message about all the special conventions in the linear algebra package. You can end the line with a colon to suppress this message.]

```
> A:=matrix([[0,-1],[1,0]]);
```

[Notice the double square brackets at the start and finish.]

```
> B:=t*A;
```

[Notice the asterisk represents scalar multiplication.]

```
> C:=exponential(B);

> det(C);

> Ct:=transpose(C);

> Ct&*C;

> evalm(%);
```

[The inclusion of the determinant of C in this sample is just for your information. The next calculations check whether $C = \exp(tA)$ is in the Euclidean orthogonal group $O(2, \mathbb{R})$ by calculating $C^T C$. Notice the strange notation &* for matrix multiplication. The command "evalm" causes Maple to simplify the entries in a matrix. The symbol % indicates that the command evalm refers to the matrix in the preceding line.]

4.4 Properties of the matrix exponential

In this section we'll concentrate on algebraic properties of the matrix exponential, postponing issues of convergence until section 9, an appendix to this chapter. For the problems in this section, again use our definition

$$\exp(A) = I_n + A + \frac{1}{2!}A^2 + \cdots + \frac{1}{m!}A^m + \cdots$$

for $A \in \mathcal{M}(n, \mathbb{R})$. Also, *assume* the following facts hold for every choice of the positive integer n and every $A, B \in \mathcal{M}(n, \mathbb{R})$:

- This series always converges; in fact, it converges absolutely. (This means $\exp(tA)$ is infinitely differentiable, and you can determine the series for its derivative by differentiating the series for $\exp(tA)$ term by term.)
- If $AB = BA$, then $\exp(A + B) = \exp(A)\exp(B)$. (If A and B *don't* commute, this might not be true.)
- You can do algebra with series much as you would with finite sums. For example,

$$P\exp(A) = P + PA + \frac{1}{2!}PA^2 + \cdots + \frac{1}{m!}PA^m + \cdots$$

  for $P, A \in \mathcal{M}(n, \mathbb{R})$, etc.

Using these facts, give informal arguments in the following problems.

Problem 4.4.1. (S)
Fix a positive integer n.

a. Show that if 0_n is the n by n zero matrix, then $\exp(0_n) = I_n$.
b. Let $A \in \mathcal{M}(n, \mathbb{R})$. Show $\exp(A)\exp(-A) = I_n$. (Hint: $A(-A) = (-A)A$.) Explain why this means $\exp(A)$ is invertible. What is $(\exp(A))^{-1}$?
c. Let $A \in \mathcal{M}(n, \mathbb{R})$. Prove by induction that $(A^T)^m = (A^m)^T$ for all positive integers m.
d. Let $A \in \mathcal{M}(n, \mathbb{R})$. Show $\exp(A^T) = (\exp(A))^T$.

Problem 4.4.2.
Fix a positive integer n, and let $A \in \mathcal{M}(n, \mathbb{R})$ and $P \in GL(n, \mathbb{R})$.

a. Prove by induction that $(PAP^{-1})^m = PA^mP^{-1}$ for all positive integers m.
b. Show $\exp(PAP^{-1}) = P\exp(A)P^{-1}$.

Problem 4.4.3.
Fix a positive integer n.

a. Choose any $t \in \mathbb{R}$ and $A \in \mathcal{M}(n, \mathbb{R})$. Prove by induction that $(tA)^m = t^m A^m$ for all positive integers m.
b. Use (a) to show

$$\exp(tA) = I_n + tA + \frac{t^2}{2!}A^2 + \cdots + \frac{t^m}{m!}A^m + \cdots$$

c. **(S)** Choose a fixed matrix $A \in \mathcal{M}(n, \mathbb{R})$ and let $\gamma(t) = \exp(tA)$. By differentiating term-by-term, show $\gamma'(t) = A\gamma(t)$. What is $\gamma'(0)$?
d. Why does (c) tell you γ is smooth?
e. Continuing the notation of (c), how do you know that $\gamma(t)$ is invertible for all $t \in \mathbb{R}$?

Problem 4.4.4.

Fix a positive integer n and a matrix $A \in \mathcal{M}(n, \mathbb{R})$. Use preceding results from this section as needed to explain why $\gamma = \exp(tA)$ is a *smooth* curve and to show that $\gamma(t)$ is a one-parameter subgroup of $GL(n, \mathbb{R})$.

One of the great theorems of the Lie theory is that for every positive integer n, *every* one-parameter subgroup of $GL(n, \mathbb{R})$ has the form $\gamma(t) = \exp(tA)$ for some $A \in \mathcal{M}(n, \mathbb{R})$.

> For any positive integer n, if $\gamma\colon \mathbb{R} \to GL(n, \mathbb{R})$ is a continuous group homomorphism, then $\gamma(t) = \exp(tA)$ for $A = \gamma'(0)$.

There is a sketch of the proof of this lovely theorem in section 9. (It is the proof Howe gives in his *Monthly* article "Very Basic Lie Theory.") The outline of the general proof has a strong resemblance to our proof for the special case when $n = 1$.

Problem 4.4.5.

Can a one-parameter subgroup "cross itself"? (This question is from *Algebra* by M. Artin.) That is, can it happen that $\gamma(t) = \gamma(s)$ for $s \neq t$ for a one-parameter subgroup $\gamma\colon \mathbb{R} \to GL(n, \mathbb{R})$?

Problem 4.4.6. (C)

Fix a positive integer n and let $A \in \mathcal{M}(n, \mathbb{R})$, and assume $A^m = 0_n$ for some positive integer m. (Such a matrix A is called *nilpotent*.) Let $B = \exp(A)$. Show that $(B - I_n)^m = 0_n$. (Such a matrix B is called *unipotent*. So the exponential map takes nilpotent matrices to unipotent matrices.)

Problem 4.4.7. (C)

(This problem comes from *Algebra* by M. Artin.) Suppose the 2 by 2 matrix A has distinct eigenvalues α and β so that

$$f(x) = \det(A - xI) = (x - \alpha)(x - \beta) \qquad \text{and} \qquad A = P \begin{bmatrix} \alpha & 0 \\ 0 & \beta \end{bmatrix} P^{-1},$$

for some invertible matrix P. *Assume* the Cayley-Hamilton theorem, which tells you that

$$f(A) = (A - \alpha I_2)(A - \beta I_2) = 0_2.$$

By the division algorithm for polynomials (i.e., long division), if m is a positive integer, then

$$x^m = f(x)q_m(x) + r_m(x),$$

where $q_m(x)$ and $r_m(x)$ are polynomials, and $\deg r_m(x) < \deg f(x) = 2$, so $\deg r_m(x) \leq 1$ and $r_m(x) = a_m x + b_m$ for real numbers a_m and b_m.

a. Show that for all nonnegative integers m, $a_m = \dfrac{\alpha^m - \beta^m}{\alpha - \beta}$ and $b_m = \dfrac{\beta^m \alpha - \alpha^m \beta}{\alpha - \beta}$.

b. Use (a) to show

$$\exp(A) = \left(\frac{e^\alpha - e^\beta}{\alpha - \beta}\right) A + \left(\frac{\alpha e^\beta - \beta e^\alpha}{\alpha - \beta}\right) I_2.$$

c. What does the result in (b) give in the special case when $P = I_2$ and A is diagonal? Compare this to what you get when you calculate $\exp(A)$ directly in this case.

4.5 Using exp to determine $L(G)$

In this section we'll exploit the properties of the matrix exponential to determine the tangent spaces of more general matrix groups. We'll also work toward the proof of a very useful theorem of the Lie theory:

$$\det(\exp(A)) = e^{tr(A)}$$

for all A in $\mathcal{M}(n, \mathbb{R})$, for any positive integer n. A nice way to express this theorem is with the following *commutative* diagram.

$$\begin{array}{ccc} GL(n,\mathbb{R}) & \stackrel{\det}{\longrightarrow} & GL(1,\mathbb{R}) \\ \exp\uparrow & & \exp\uparrow \\ \mathcal{M}(n,\mathbb{R}) & \stackrel{\mathrm{tr}}{\longrightarrow} & \mathcal{M}(1,\mathbb{R}) \end{array}$$

Saying the diagram is commutative means that you get the same result if you follow either set of arrows from $\mathcal{M}(n, \mathbb{R})$.

Although we won't explore this, it's important to note that the exponential function is much more than a tool for producing one-parameter subgroups to help us describe tangent spaces. There is also a series definition of a matrix version of the natural logarithm. The important fact is that the functions exp and log provide a "local" one-to-one and onto correspondence between a "neighborhood" of the zero matrix in the vector space $\mathcal{M}(n, \mathbb{R})$ and a neighborhood of the identity matrix in the group $GL(n, \mathbb{R})$. (This correspondence is made more precise in section 9.) This correspondence helps to illuminate the heart of the Lie theory: studying the tangent space of a Lie group tells us something important about the group itself. See section 7 for the series defining the log and some of its properties. We'll return to these ideas in Chapter 5.

Problem 4.5.1.

Fix a positive integer n. If $D \in \mathcal{M}(n, \mathbb{R})$ is a diagonal matrix with entries $d_1, d_2, \ldots, d_n$ down the main diagonal (and all other entries equal to zero), write $D = \mathrm{diag}\,(d_1, \ldots, d_n)$. Assume that $A = \mathrm{diag}\,(a_1, \ldots, a_n)$ and $B = \mathrm{diag}\,(b_1, \ldots, b_n)$ implies $AB = \mathrm{diag}\,(a_1 b_1, \ldots, a_n b_n)$ and $\det(A) = a_1 \cdots a_n$.

a. Use induction to show that $D^m = \mathrm{diag}\,(d_1^m, \ldots, d_n^m)$ for all integers $m \geq 0$.

b. Use (a) to show that $\exp(D)$ is also a diagonal matrix, $\exp(D) = \mathrm{diag}\,(e^{d_1}, e^{d_2}, \ldots, e^{d_n})$.

c. Show that $\det(\exp(D)) = e^{tr(D)}$ in this case.

Problem 4.5.2.

Fix $n > 0$. Assume $A \in \mathcal{M}(n, \mathbb{R})$, and $A = PDP^{-1}$ for a diagonal matrix D and an invertible matrix P, where D and P are also n by n matrices. Show that $\det(\exp(A)) = e^{tr(A)}$ in this case.

Problem 4.5.3.

Let $G = \mathcal{L}(2, \mathbb{R})$, the Lorentz group, and let C be the matrix describing the Lorentz dot product, as usual. Let

$$W = \left\{ \begin{bmatrix} 0 & b \\ b & 0 \end{bmatrix} : b \in \mathbb{R} \right\}.$$

By (3.2.1) we know that W is a subspace of $\mathcal{M}(2, \mathbb{R})$. (In particular, $B \in W$ implies $tB \in W$ for all $t \in \mathbb{R}$.)

a. (S) Assume $\gamma\colon \mathbb{R} \to \mathcal{L}(2, \mathbb{R})$ is a smooth curve passing through the identity. Differentiate both sides of the equation $\gamma(t)^T C \gamma(t) = C$ with respect to t and then evaluate at $t = 0$ to show $\gamma'(0) \in W$. Explain why this shows $L(G) \subseteq W$.

b. (S) Show that $B \in W$ implies $\exp(tB) \in G$.

c. Explain why (c) shows that every matrix in W is contained in $L(G)$, $W \subseteq L(G)$, and therefore completes the proof that $L(G) = W$.

d. Show that $\det(\exp(B)) = e^{tr(B)}$ in this case.

Problem 4.5.4.

Fix a positive integer n, and let $G = SL(n, \mathbb{R})$, the special linear group. Problem (3.1.11) showed that if $\gamma\colon \mathbb{R} \to SL(2, \mathbb{R})$ is a smooth curve passing through the identity, then $\mathrm{tr}(\gamma'(0)) = 0$. The argument involved differentiating both sides of the equation $\det(\gamma(t)) = 1$ with respect to t and then evaluating at $t = 0$. Exactly the same argument works for arbitrary n; assume that this is so.

a. Explain why this assumption tells you that

$$L(G) \subseteq W = \{B \in \mathcal{M}(n, \mathbb{R}) : \mathrm{tr}(B) = 0\}.$$

b. (S) Show W is a subspace of $\mathcal{M}(n, \mathbb{R})$. (In particular, $B \in W$ implies $tB \in W$ for all $t \in \mathbb{R}$.)

c. *Assume* $\det(\exp(B)) = e^{tr(B)}$ in this case, and use this fact to show that $\exp(tB) \in G$ for every $B \in W$.

d. Explain why the result of (c) tells you that $W \subseteq L(G)$ and so completes the proof that $L(G) = W$.

Problem 4.5.5.

Fix a positive integer n and let $G = O(n, \mathbb{R})$, the Euclidean orthogonal group. Problem (3.1.12) actually shows that for all positive integers n (not just $n = 2$), if $\gamma\colon \mathbb{R} \to O(n, \mathbb{R})$ is a smooth curve passing through the identity, then $\gamma'(0) + (\gamma'(0))^T = 0_n$. Assume that this is so.

a. Explain why this assumption tells you that

$$L(G) \subseteq W = \{B \in \mathcal{M}(n, \mathbb{R}) : B + B^T = 0_n\}.$$

b. Show W is a subspace of $\mathcal{M}(n, \mathbb{R})$. (In particular, $B \in W$ implies $tB \in W$ for all $t \in \mathbb{R}$.)

c. Show that $B \in W$ implies $\exp(tB) \in G$. (Hint: $B^T = -B$ implies $BB^T = B^T B$.)

d. Explain why the result of (c) tells you that $W \subseteq L(G)$ and so completes the proof that $L(G) = W$.

e. *Assume* $\det(\exp(B)) = e^{tr(B)}$ in this case, and use this fact to show that $\exp(tB)$ is actually in $SO(n, \mathbb{R})$ for every $B \in W$.

f. Explain why $O(n, \mathbb{R})$ and $SO(n, \mathbb{R})$ have the same tangent space at the identity.

Problem 4.5.6.

Fix a positive integer n. The goal of this problem is to prove $\det(\exp(A)) = e^{tr(A)}$ in the general case where A is *any* matrix in $\mathcal{M}(n, \mathbb{R})$. To do so, we make use of three facts. (Assume them, whether or not you've done the problems leading to their proofs.)

(i) The tangent space at the identity of $GL(n, \mathbb{R})$ is $\mathcal{M}(n, \mathbb{R})$.

(ii) The differential of the determinant map is the trace. In other words, if $F(A) = \det(A)$, so $F: GL(n, \mathbb{R}) \to GL(1, \mathbb{R})$ and $dF: \mathcal{M}(n, \mathbb{R}) \to \mathcal{M}(1, \mathbb{R})$, then $dF(A) = \operatorname{tr}(A)$.

(iii) Every one-parameter subgroup of $GL(1, \mathbb{R})$ has the form $\exp(ta)$ for some $a \in \mathbb{R}$. (The group $\mathcal{M}(1, \mathbb{R})$ under matrix addition is isomorphic to the group $\mathbb{R}$ under addition; identify them. Similarly, identify $GL(1, \mathbb{R})$ with the isomorphic group of nonzero real numbers under multiplication.)

Fix $A \in \mathcal{M}(n, \mathbb{R})$. Define $\alpha(t) = \exp(tA)$.

a. Explain why α is a one-parameter subgroup of $GL(n, \mathbb{R})$.

b. Define $\gamma: \mathbb{R} \to \mathbb{R}$ by

$$\gamma(t) = \det(\alpha(t)) = \det(\exp(tA)).$$

Explain why γ is a one-parameter subgroup of $GL(1, \mathbb{R})$.

c. Explain why $\gamma'(0) = \operatorname{tr}(\alpha'(0))$.

d. Explain why (c) implies $\det(\exp(tA)) = e^{tr(A)t}$ (and $t = 1$ gives our result).

Both the orthogonal group and the symplectic group have been described as symmetry groups preserving a non-degenerate bilinear form, symmetric in the case of the orthogonal group and anti-symmetric in the case of the symplectic group. In the next problem you'll determine the tangent spaces of both groups. (Compare (3.1.16).)

Problem 4.5.7. (C)

Fix a positive integer n.

a. Fix $C \in GL(n, \mathbb{R})$ and let $G = \{M \in \mathcal{M}(n, \mathbb{R}) : M^T C M = C\}$. You can assume G is a group under matrix multiplication.

Let $W = \{A \in \mathcal{M}(n, \mathbb{R}) : A^T C + CA = 0_n\}$. Show that $L(G) = W$.

b. Suppose $C = I_n$, the case of the Euclidean orthogonal group $O(n, \mathbb{R})$. What is the dimension of $L(G)$?

c. Suppose $n = 2m$ and

$$C = \begin{bmatrix} 0_m & I_m \\ -I_m & 0_m \end{bmatrix}.$$

What is the dimension of $L(G)$ in terms of m? (Hint: For $M \in L(G)$, write

$$M = \begin{bmatrix} M_{11} & M_{12} \\ M_{21} & M_{22} \end{bmatrix},$$

where the M_{ij} are m by m matrices.)

Problem 4.5.8.

Fix a positive integer n. In this problem you'll find $L(G)$ for G the subgroup of $GL(n, \mathbb{R})$ consisting of diagonal matrices with positive entries on the diagonal, $G = \{\text{diag}(e^{a_1}, \ldots, e^{a_n}) : a_j \in \mathbb{R}\}$.

a. Show that G is an abelian group under matrix multiplication.

b. Show that G is isomorphic to $\mathbb{R}^n$ viewed as a group under addition.

c. Determine $L(G)$.

4.6 Differential equations

From calculus we know that the solution to the initial value problem

$$y' = \frac{dy}{dt} = ay \quad y(0) = x$$

is $y = y(t) = xe^{ta}$. (It may look strange to call the initial value x, but the choice is motivated by thinking of the solution y as a function both of time t and of position x. We'll have more to say about this later.) The matrix exponential gives us the following lovely generalization for any positive integer n.

> Choose A in $\mathcal{M}(n, \mathbb{R})$, and let $\boldsymbol{x} = (x_1, \ldots, x_n)$ be a fixed vector in $\mathbb{R}^n$. Suppose $\boldsymbol{y} = (y_1, \ldots, y_n)$ where the $y_j \colon \mathbb{R} \to \mathbb{R}$ are functions of a variable t. Then the solution of the initial value problem $\boldsymbol{y}' = A\boldsymbol{y}$ and $\boldsymbol{y}(0) = \boldsymbol{x}$ is $\boldsymbol{y} = \exp(tA)\boldsymbol{x}$.

Problem 4.6.1. (S)

Verify that $\boldsymbol{y} = \exp(tA)\boldsymbol{x}$ is indeed a solution of the initial value problem

$$\frac{d\boldsymbol{y}}{dt} = \boldsymbol{y}' = A\boldsymbol{y} \qquad \text{and} \qquad \boldsymbol{y}(0) = \boldsymbol{x}$$

by differentiating $\exp(tA)\boldsymbol{x}$ with respect to t and by evaluating $\exp(tA)\boldsymbol{x}$ at $t = 0$. (Assume that the product rule holds in differentiating $\exp(tA)\boldsymbol{x}$.)

The vector notation somewhat hides the fact that $\boldsymbol{y}' = A\boldsymbol{y}$ is actually a *system* of n differential equations, one for each function $y_j = y_j(t)$, along with n initial conditions $y_j(0) = x_j$. Since each differential equation has the form

> y_j' equals a *linear combination* of the functions $y_1, \ldots, y_n$,
> with constant coefficients a_{ij} and constant term 0,

it is a system of *linear homogeneous* differential equations.

In the problems that follow, you can use your results from section 4.3, or you can use a computer algebra system to recalculate them as needed. (You definitely should use the computer for (4.6.11a).) To make typing easier, the derivative with respect to t is denoted by a prime, instead of dy/dt, etc.

Problem 4.6.2. (S)

a. Use the matrix exponential to solve the following initial value problem.

$$\begin{aligned} y_1' &= y_1 & y_1(0) &= x_1 \\ y_2' &= -y_2 & y_2(0) &= x_2 \end{aligned}$$

b. Use calculus to check your answer.

Problem 4.6.3.

a. Use the matrix exponential to solve the following initial value problem.

$$\begin{aligned} y_1' &= -y_2 & y_1(0) &= x_1 \\ y_2' &= y_1 & y_2(0) &= x_2 \end{aligned}$$

b. Use calculus to check your answer.

Problem 4.6.4.

a. Use the matrix exponential to solve the following initial value problem.

$$\begin{aligned} y_1' &= y_3 & y_1(0) &= x_1 \\ y_2' &= 0 & y_2(0) &= x_2 \\ y_3' &= -y_1 & y_3(0) &= x_3 \end{aligned}$$

b. Use calculus to check your answer.

Problem 4.6.5.

In this problem

$$A = \begin{bmatrix} 0 & 1 \\ 1 & 0 \end{bmatrix}.$$

a. Write down the system of differential equations and initial conditions corresponding to $\mathbf{y}' = A\mathbf{y}$ and $\mathbf{y}(0) = \mathbf{x}$.

b. Use the matrix exponential to solve the initial value problem.

c. Use calculus to check your answer.

Problem 4.6.6.

In this problem

$$A = \begin{bmatrix} 0 & -1 & 0 \\ 1 & 0 & 0 \\ 0 & 0 & 0 \end{bmatrix}.$$

a. Write down the system of differential equations and initial conditions corresponding to $\mathbf{y}' = A\mathbf{y}$ and $\mathbf{y}(0) = \mathbf{x}$.

b. Use the matrix exponential to solve the initial value problem.

c. Use calculus to check your answer.

Problem 4.6.7.

For this problem,

$$A = \begin{bmatrix} 0 & 1 & 0 & 0 \\ 1 & 0 & 0 & 0 \\ 0 & 0 & 0 & 0 \\ 0 & 0 & 0 & 0 \end{bmatrix}.$$

a. Write down the system of differential equations and initial conditions corresponding to $\boldsymbol{y}' = A\boldsymbol{y}$ and $\boldsymbol{y}(0) = \boldsymbol{x}$.

b. Use the matrix exponential to solve the initial value problem.

c. Use calculus to check your answer.

Problem 4.6.8.

a. Use methods from Calculus I to solve the following initial value problem.

$$\begin{aligned} s' &= v & s(0) &= s_0 \\ v' &= a & v(0) &= v_0 \\ a' &= 0 & a(0) &= g \quad (g \text{ is a constant}) \end{aligned}$$

b. Now use the matrix exponential to solve it.

Problem 4.6.9.

a. **(S)** Use the matrix exponential to solve the following initial value problem.

$$\begin{aligned} y_1' &= y_1 + 3y_2 & y_1(0) &= x_1 \\ y_2' &= 3y_1 + y_2 & y_2(0) &= x_2 \end{aligned}$$

b. Use calculus to check your answer to (a).

c. Solve this problem using (4.4.7b) instead.

Problem 4.6.10.

a. Use the matrix exponential to solve the following initial value problem.

$$\begin{aligned} y_1' &= y_1 + y_2 & y_1(0) &= x_1 \\ y_2' &= y_2 & y_2(0) &= x_2 \end{aligned}$$

b. Use calculus to check your answer to (a).

Problem 4.6.11.

a. Use the matrix exponential to solve the following initial value problem.

$$\begin{aligned} y_1' &= -29y_1 + 24y_2 - 76y_3 & y_1(0) &= x_1 \\ y_2' &= -12y_1 + 11y_2 - 30y_3 & y_2(0) &= x_2 \\ y_3' &= 8y_1 - 6y_2 + 22y_3 & y_3(0) &= x_3 \end{aligned}$$

b. Use calculus to check your answer to (a). (This is messy, but straightforward.)

For a fixed positive integer n and a fixed $A \in \mathcal{M}(n, \mathbb{R})$, let $F(t, \boldsymbol{x}) = \exp(tA)\boldsymbol{x}$ for $t \in \mathbb{R}$ and $\boldsymbol{x} \in \mathbb{R}^n$. We'll think of t as time and $\boldsymbol{x}$ as position in $\mathbb{R}^n$. There is a geometric interpretation of these ideas that explains the use of the term "flow" for the function F. Also, it is this function-of-two-variables point of view that motivates the use of the notation $\boldsymbol{x}$ for the vector of initial conditions.

To explain the geometry, first we need the idea of a **vector field**. A vector field is a function $f: \mathbb{R}^n \to \mathbb{R}^m$. When $n = m = 2$ (or $n = m = 3$) we can visualize a vector field as follows. Regard the input vector $\boldsymbol{y}$ as a point in $\mathbb{R}^2$. Draw the output vector $f(\boldsymbol{y})$ as an arrow beginning at the point $\boldsymbol{y}$. This is the same convention we use to draw tangent vectors. In fact, we can actually regard the vectors $f(\boldsymbol{y})$ as tangent vectors, but first we have to specify what they're tangent *to*.

Suppose we have a system of differential equations $\boldsymbol{y}' = f(\boldsymbol{y})$. Ordinarily the system has many different solutions, depending on the choice of the initial condition $\boldsymbol{y}(0) = \boldsymbol{x}$. The vectors in the vector field are "direction vectors" for the various solutions. Indeed the vectors $f(\boldsymbol{y})$ are tangent to the solution curves.

Our methods apply to the special case where f is a linear transformation, $\boldsymbol{y}' = f(\boldsymbol{y}) = A\boldsymbol{y}$. Here's an example from the *Monthly* article "Very Basic Lie Theory" by Howe, in slightly different notation. Let $\boldsymbol{y} = (y_1, y_2)$ and consider the initial value problem

$$\frac{d\boldsymbol{y}}{dt} = f(\boldsymbol{y}) = (-y_2, y_1) \quad \boldsymbol{y}(0) = (x_1, x_2) = \boldsymbol{x}.$$

As above, we regard $f(\boldsymbol{y})$ as a vector field; that is, we interpret f as a function that assigns to a point (y_1, y_2) in the plane $\mathbb{R}^2$ the vector with coordinates $(-y_2, y_1)$. We can draw a picture of some of the vectors in this vector field, using the convention that the vector $f(\boldsymbol{y})$ is drawn beginning from the point (y_1, y_2). Here are some sample values we will use.

(y_1, y_2)	$(1, 0)$	$(1, 1,)$	$(0, 1)$	$(-1, 1)$	$(-1, 0)$	$(-1, -1)$	$(0, -1)$	$(1, -1)$
$f(\boldsymbol{y})$	$(0, 1)$	$(-1, 1)$	$(-1, 0)$	$(-1, -1)$	$(0, -1)$	$(1, -1)$	$(1, 0)$	$(1, 1)$

Below is a sketch of the vector field $f(y_1, y_2) = (-y_2, y_1)$ corresponding to the sample values above.

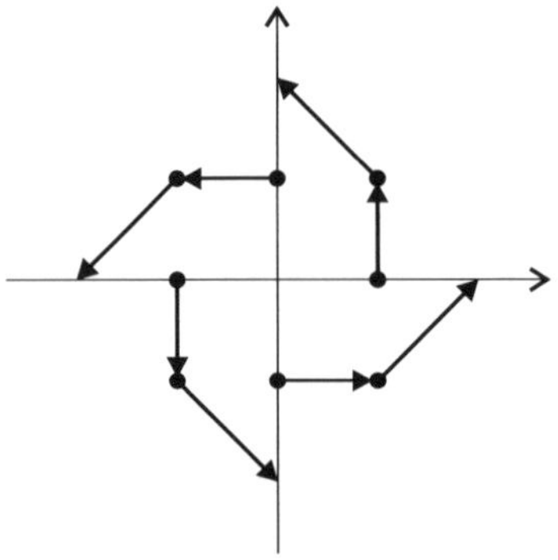

Figure 4.1. Sketch of the vector field $f(y_1, y_2) = (-y_2, y_1)$

Now we return to the differential equation. Notice that it can be rewritten in matrix notation as

$$\frac{d\boldsymbol{y}}{dt} = \begin{bmatrix} y_1' \\ y_2' \end{bmatrix} = \begin{bmatrix} 0 & -1 \\ 1 & 0 \end{bmatrix} \begin{bmatrix} y_1 \\ y_2 \end{bmatrix}.$$

We are looking for a solution $\boldsymbol{y} = \exp(tA)\boldsymbol{x}$. We can see that we should choose

$$A = \begin{bmatrix} 0 & -1 \\ 1 & 0 \end{bmatrix}.$$

We have previously calculated that

$$\exp(tA) = \begin{bmatrix} \cos(t) & -\sin(t) \\ \sin(t) & \cos(t) \end{bmatrix},$$

so our solution is

$$\begin{aligned} y_1 &= x_1\cos(t) - x_2\sin(t), \\ y_2 &= x_1\sin(t) + x_2\cos(t). \end{aligned}$$

A slight change of notation makes it easier to interpret the curve

$$t \mapsto \boldsymbol{y}(t) = \exp(tA)\boldsymbol{x} = F(t, \boldsymbol{x})$$

in this case. Let $a = |\boldsymbol{x}| = \sqrt{x_1^2 + x_2^2}$, and choose b so that $\boldsymbol{x} = (x_1, x_2) = (a\cos(b), a\sin(b))$; in other words, switch to polar coordinates. Then the solution $\boldsymbol{y}$ can be written

$$\begin{aligned} y_1 &= a\cos(b)\cos(t) - a\sin(b)\sin(t), \\ y_2 &= a\cos(b)\sin(t) + a\sin(b)\cos(t). \end{aligned}$$

We can rewrite this as

$$\boldsymbol{y} = (y_1, y_2) = (a\cos(t+b), a\sin(t+b)).$$

For a fixed initial value $\boldsymbol{x}$, we can now see that the solution is a circle of radius $a = |\boldsymbol{x}|$ centered at the origin, and the curve $t \mapsto \boldsymbol{y}(t)$ traces the circle out counter-clockwise, starting at $\boldsymbol{x}$ when $t = 0$. The corresponding flow is the set of concentric circles, one for each choice of $\boldsymbol{x}$. (See Figure 4.2.) The flow is a picture of the function $F(t, \boldsymbol{x}) = \exp(tA)\boldsymbol{x}$. Compare this flow to the sketch of the vector field given by $f(\boldsymbol{y})$ in Figure 4.1.

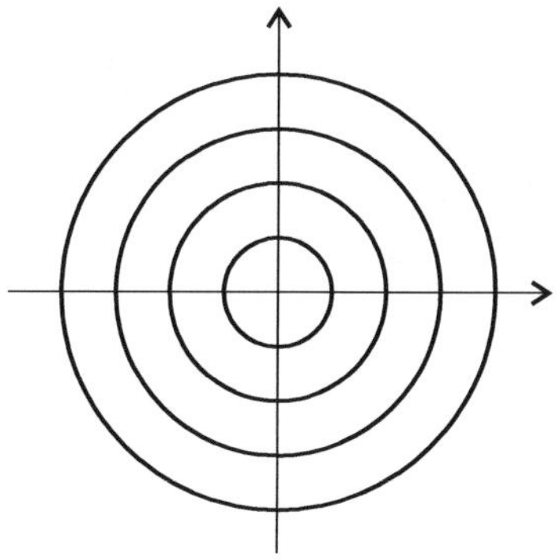

Figure 4.2. The flow associated with the vector field $f(y_1, y_2) = (-y_2, y_1)$

Problem 4.6.12.

Carry out an analysis similar to the one above for the system

$$\frac{d\mathbf{y}}{dt} = f(\mathbf{y}) = (y_2, y_1) \quad \mathbf{y}(0) = \mathbf{x} = (x_1, x_2)$$

as follows

a. Sketch the vector field $f(\mathbf{y})$. Include vectors in all four quadrants; also include vectors for $y_1 = \pm y_2$.

b. Solve this system of differential equations: find formulas for y_1 and y_2 as functions of t.

c. Take advantage of identities for cosh and sinh to rewrite your formulas in (b) in a way that makes it easier to interpret the curves $t \mapsto \mathbf{y}(t)$. Sketch some sample curves on the same axes with your vector field sketch. Include examples with $|x_1| < |x_2|$, with $|x_2| < |x_1|$ and with $|x_1| = |x_2|$.

You may be wondering what happens with a non-linear vector field. Here is a picture of $f(\mathbf{y}) = f(y_1, y_2) = (y_1 y_2, y_1 + y_2)$ and some corresponding solution curves to the initial value problem $\mathbf{y}' = f(\mathbf{y})$ and $\mathbf{y}(0) = \mathbf{x}$.

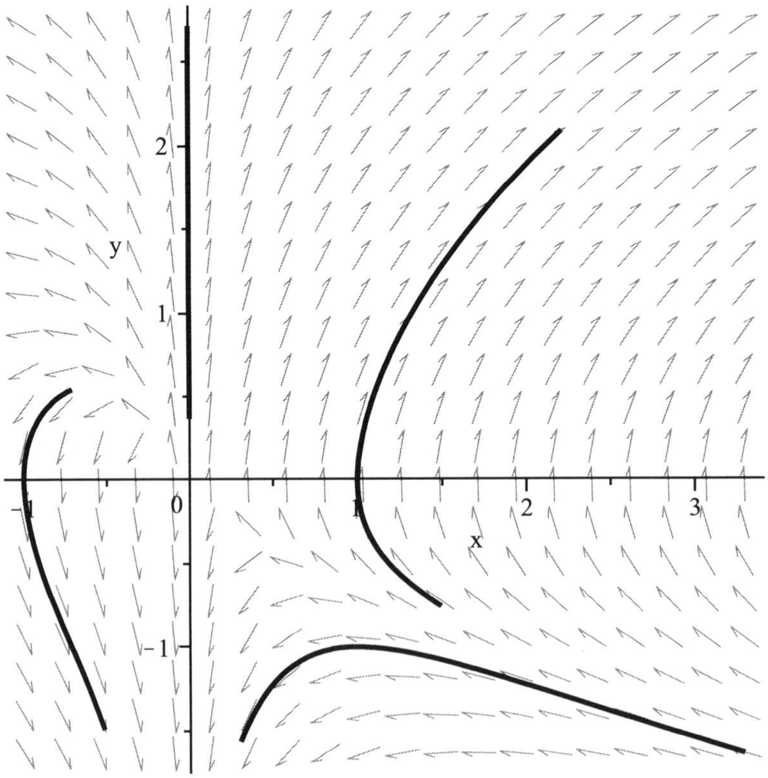

Figure 4.3. The vector field $f(\mathbf{y}) = f(y_1, y_2) = (y_1 y_2, y_1 + y_2)$ and some solution curves for $\mathbf{y}' = f(\mathbf{y})$, $\mathbf{y}(0) = \mathbf{x}$

4.7 Putting the pieces together

Definition Fix a positive integer n. A **one-parameter subgroup** of a matrix group $G \subseteq GL(n, \mathbb{R})$ is (the image of) a continuous group homomorphism $\gamma: \mathbb{R} \to G$.

Here $\mathbb{R}$ is a group under addition, and G is a group under matrix multiplication, so the requirement that γ be a group homomorphism means

$$\gamma(t+s) = \gamma(t)\gamma(s) \qquad \text{for all } t, s \in \mathbb{R}.$$

Note that every one-parameter subgroup of a group G contained in $GL(n, \mathbb{R})$ satisfies $\gamma(0) = I_n$.

The key tool for our analysis of one-parameter subgroups is the **exponential function** for matrices:

Definition Fix a positive integer n. For A in $\mathcal{M}(n, \mathbb{R})$, define

$$\exp(A) = I_n + A + \frac{A^2}{2!} + \frac{A^3}{3!} + \cdots + \frac{A^m}{m!} + \cdots.$$

A discussion of the *convergence* of this series is postponed to the appendix to this chapter, section 9.

Proposition 4.1 Algebraic properties of the exponential function. *Fix a positive integer n and let $A, B \in \mathcal{M}(n, \mathbb{R})$.*

a. *We have* $\exp(0_n) = I_n$.

b. *If $AB = BA$, then* $\exp(A + B) = \exp(A)\exp(B)$.

c. *For all A,* $\exp(A)^{-1} = \exp(-A)$.

d. *For all A,* $\exp(A^T) = (\exp(A))^T$.

e. *For all A and all $P \in GL(n, \mathbb{R})$,* $\exp(PAP^{-1}) = P\exp(A)P^{-1}$.

Proposition 4.2 *Fix a positive integer n and a matrix $A \in \mathcal{M}(n, \mathbb{R})$ and let $\gamma(t) =$* $\exp(tA)$ *for $t \in \mathbb{R}$. Then:*

a. *The function $\gamma(t)$ is differentiable (in fact, it is smooth), and $\gamma'(t)$ equals the sum of the series obtained by differentiating the series for $\gamma(t)$ term by term. Further, $\gamma'(t) = A\gamma(t)$, so $\gamma'(0) = A$.*

b. *The function $\gamma: \mathbb{R} \to GL(n, \mathbb{R})$ is a one-parameter subgroup.*

Theorem 4.3

a. ***The case n = 1.*** *If γ is a one-parameter subgroup,*

$$\gamma: \mathbb{R} \to GL(1, \mathbb{R}),$$

then γ has the form $\gamma(t) = \exp(at)$, for all $t \in \mathbb{R}$, where $a = \gamma'(0)$. (As usual, $\exp(t)$ means e^t.) In particular, γ is smooth.

b. ***The case when n is any positive integer.*** *If*

$$\gamma\colon \mathbb{R} \to GL(n, \mathbb{R})$$

is a one-parameter subgroup, then

$$\gamma(t) = \exp(tA),$$

where $A = \gamma'(0)$. *In particular,* γ *is smooth.*

A sketch of the proof of this important theorem for the case $n > 1$ appears in section 9. Because there is no requirement that one-parameter subgroups be *onto* maps, Theorem 4.3 applies to any subgroup of $GL(n, \mathbb{R})$:

Corollary to Theorem 4.3 *Suppose* n *is a positive integer and* G *is a subgroup of* $GL(n, \mathbb{R})$. *If*

$$\gamma\colon \mathbb{R} \to G$$

is a one-parameter subgroup, then

$$\gamma(t) = \exp(tA),$$

where $A = \gamma'(0) \in L(G)$.

As noted in the discussion of Hilbert's fifth problem in Chapter 3, an even more general theorem than Theorem 4.3 is true, but proving it is well beyond what our methods can accomplish:

Theorem 4.3* ***The general case.*** *If* G_1 *and* G_2 *are Lie groups and*

$$F\colon G_1 \to G_2$$

is a continuous group homomorphism, then F *is smooth and hence a Lie group homomorphism.*

It's seductive to think that the corollary to Theorem 4.3 tells us that $A \in L(G)$ implies $\exp(A) \in G$, but it does not. If $A \in L(G)$, then we do know $A = \gamma'(0)$ for a smooth curve $\gamma\colon \mathbb{R} \to G$ passing through the identity. However we can't assume that γ is a homomorphism, and thus we can't apply Theorem 4.3 and its corollary. In fact, the statement is true, but using our relatively elementary methods, we are able to prove it only for specific matrix groups. We state it as Theorem 4.4*.

Theorem 4.4* *Fix a positive integer* n *and assume* G *is a Lie subgroup of* $GL(n, \mathbb{R})$ *and its tangent space at the identity is* $L(G)$. *Then* $A \in L(G)$ *implies* $\exp(A) \in G$, *and, further,* $\gamma(t) = \exp(tA)$ *is a one-parameter subgroup* $\gamma\colon \mathbb{R} \to G$ *with* $A = \gamma'(0)$.

So far, we have verified Theorem 4.4* for the cases where G is $GL(n, \mathbb{R})$, the special linear group $SL(n, \mathbb{R})$, the Lorentz group $\mathcal{L}(2, \mathbb{R})$, the group $\mathcal{L}(4, \mathbb{R})$ of symmetries of four-dimensional spacetime, and the Euclidean orthogonal group $O(n, \mathbb{R})$. In the next chapter, we will verify Theorem 4.4* for the general orthogonal group, the symplectic group, the *Heisenberg group*, and the group of diagonal matrices with positive entries.

Remark Some authors build the truth of Theorem 4.4* into their *definition* of the tangent space for $G \subseteq GL(n, \mathbb{R})$ by defining $L(G) = \{A \in \mathcal{M}(n, \mathbb{R}) : \exp(tA) \in G \text{ for all } t \in \mathbb{R}\}$. Our definition is equivalent. (See, for example, Curtis' *Matrix Groups*, chapter 12, section B, for the correspondence between the tangent space at the identity and the "left-invariant vector fields" of the formal definition of the Lie algebra of a general Lie group.)

In addition to its role in the Lie theory, the exponential function for matrices gives us solutions to *linear homogeneous* systems of differential equations.

Notation Fix a positive integer n. Suppose $y_1 = y_1(t), \ldots, y_n = y_n(t)$ are differentiable functions of the real variable t. Write $\boldsymbol{y} = (y_1, \ldots, y_n)$ and $d\boldsymbol{y}/dt = \boldsymbol{y}' = (y_1', \ldots, y_n')$. (Throughout we use the prime notation to indicate the derivative with respect to t.) Assume the a_{ij} below are constants. Then the system of differential equations

$$\begin{aligned} y_1' &= a_{11}y_1 + \ldots + a_{1n}y_n \\ \vdots &= \quad\quad \vdots \\ y_n' &= a_{n1}y_1 + \ldots + a_{nn}y_n \end{aligned}$$

is a *linear homogenous* system. It is summarized by $\boldsymbol{y}' = A\boldsymbol{y}$, where A is the n by n matrix with entries a_{ij}. Similarly, the initial conditions

$$\begin{aligned} y_1(0) &= x_1 \\ \vdots &= \vdots \\ y_n(0) &= x_n \end{aligned}$$

are summarized by $\boldsymbol{y}(0) = \boldsymbol{x}$, where $\boldsymbol{x}$ is the vector $(x_1, \ldots, x_n)$.

Proposition 4.5 *With notation as above, the solution of the initial value problem* $d\boldsymbol{y}/dt = \boldsymbol{y}' = A\boldsymbol{y}$ *and* $\boldsymbol{y}(0) = \boldsymbol{x}$ *is* $\boldsymbol{y} = \exp(tA)\boldsymbol{x}$.

Theorem 4.6

a. ***The case n = 1.*** *If γ is a differentiable group homomorphism,*

$$\gamma\colon \mathbb{R} \to GL(1, \mathbb{R}),$$

then γ has the form $\gamma(t) = \exp(at)$ for all $t \in \mathbb{R}$, where $a = \gamma'(0)$.

b. ***The case when n is any positive integer.*** *If*

$$\gamma\colon \mathbb{R} \to GL(n, \mathbb{R})$$

is a differentiable *group homomorphism, then*

$$\gamma(t) = \exp(tA),$$

where $A = \gamma'(0)$.

The proof of Theorem 4.6 depends on a theorem about the *uniqueness* of solutions to the initial value problem in Proposition 4.5. Let $\gamma\colon \mathbb{R} \to GL(n, \mathbb{R})$ be a differentiable group homomorphism. If we choose $A = \gamma'(0)$, we can show that $\gamma'(t) = A\gamma(t)$. Therefore $\boldsymbol{y} = \gamma(t)\boldsymbol{x}$ is a solution to the initial value problem. We also know that $\boldsymbol{y} = \exp(tA)\boldsymbol{x}$ is a solution. It follows that $\gamma(t) = \exp(tA)$ for every real number t. □

Theorem 4.7 ***The special case of the determinant map.*** *Fix a positive integer n. For every $A \in \mathcal{M}(n, \mathbb{R})$,*

$$det(\exp(A)) = \exp(tr(A)).$$

We can express the conclusion of Theorem 4.7 by saying the following diagram is commutative—*that is, the same results are obtained by following either set of arrows from $\mathcal{M}(n, \mathbb{R})$.*

$$\begin{array}{ccc} GL(n, \mathbb{R}) & \stackrel{\text{det}}{\longrightarrow} & GL(1, \mathbb{R}) \\ \exp \uparrow & & \exp \uparrow \\ \mathcal{M}(n, \mathbb{R}) & \stackrel{\text{tr}}{\longrightarrow} & \mathcal{M}(1, \mathbb{R}) \end{array}$$

Theorem 4.7 is a special case of another major result in the Lie theory. It is more than we can prove with our methods, except for the special case of the determinant map.

Theorem 4.7* ***The general case.*** *Suppose G_1 and G_2 are Lie groups and*

$$F: G_1 \to G_2$$

is a Lie group homomorphism. Let dF be the differential of F, so dF is a linear transformation, and

$$dF: L(G_1) \to L(G_2).$$

Then the functions F, dF and exp are related as follows:

$$F(\exp(A)) = \exp(dF(A))$$

for every $A \in L(G_1)$.

The corresponding commutative diagram appears below.

$$\begin{array}{ccc} G_1 & \stackrel{F}{\longrightarrow} & G_2 \\ \exp \uparrow & & \exp \uparrow \\ L(G_1) & \stackrel{dF}{\longrightarrow} & L(G_2) \end{array}$$

Theorem 4.7 is the special case of 4.7* where the function F is the determinant map. Recall that we know from Chapter 3 that the differential of the determinant map is the trace map.

There is also a matrix version of the natural logarithm. The logarithm and exponential functions set up the "local" one-to-one correspondence between a neighborhood of the zero matrix in the tangent space and a neighborhood of the identity matrix in the Lie group that is at the heart of the Lie theory. Even though the problems in this chapter haven't explored the matrix logarithm, for completeness we include the definition here, along with some of its properties.

Definition Fix a positive integer n. For M in $GL(n, \mathbb{R})$, define the natural **logarithm** of M by

$$\log(M) = (M - I_n) - \frac{1}{2}(M - I_n)^2 + \frac{1}{3}(M - I_n)^3 - \cdots + \frac{(-1)^{m-1}}{m}(M - I_n)^m + \cdots.$$

Proposition 4.8 ***Properties of the logarithm function.*** *Fix a positive integer n. For $M \in GL(n, \mathbb{R})$ and $A \in \mathcal{M}(n, \mathbb{R})$, we have the following.*

a. *The series for $\log(M)$ converges for $\|M - I_n\| < 1$.*

b. *When A is near 0_n and M is near I_n and both the log and the exponential functions are defined,*

$$\log(\exp(A)) = A \quad \textit{and} \quad \exp(\log(M)) = M.$$

c. *If M_1 and M_2 are near I_n and $\log(M_1)$ and $\log(M_2)$ commute, then*

$$\log(M_1 M_2) = \log(M_1) + \log(M_2).$$

The proof of Proposition 4.8 (a) is in section 9. The proof of (b) is a direct, but messy, calculation from the series for log and exp. The proof of (c) follows from

$$\begin{aligned}\exp(\log(M_1 M_2)) &= M_1 M_2 \\ &= \exp(\log(M_1)) \exp(\log(M_2)) \\ &= \exp(\log(M_1) + \log(M_2)),\end{aligned}$$

and the fact that exp is invertible, and therefore one-to-one, near $0_n \in \mathcal{M}(n, \mathbb{R})$. □

Suggestions for further reading

Duzhin, S.V. and Chebotarevsky, B.D., *Transformation Groups for Beginners*, AMS Student Mathematical Library, Volume 25 (2004). Chapter 7 is on symmetries of differential equations.

Stillwell, J., *Naive Lie Theory*, Undergraduate Texts in Mathematics, Springer (2008). Chapter 7 on the matrix logarithm includes an elementary proof of the Campbell-Baker-Hausdorff (CBH) theorem: if $\exp(X)\exp(Y) = \exp(Z)$, then Z can be expressed in terms of X, Y and "brackets" of X and Y. In other words, the CBH theorem says that in a neighborhood of the identity, the group operation is determined by the bracket operation in its tangent space. We'll define the bracket operation in Chapter 5.

Tapp, K., *Matrix Groups for Undergraduates*, Student Mathematical Library, American Mathematical Society (2005). Chapter 6 on matrix exponentiation uses a slightly different approach to some ideas in this chapter.

4.8 A broader view: Lie and differential equations

Lie's study of what we now call Lie groups was motivated in large part by Galois' theory of groups of symmetries of the zeroes of polynomial equations $f(x) = 0$. For a simple example, consider the polynomial equation $x^2 + 1 = 0$ with zeroes i and $-i$ in the complex numbers $\mathbb{C}$. There are two symmetries of this equation, each an invertible function $\mathbb{C} \to \mathbb{C}$. One is the identity function $I(z) = z$ for all $z \in \mathbb{C}$. The other is the function $S(a + bi) = a - bi$ taking a complex number to its conjugate. Notice that I fixes both zeroes, S interchanges them, and the two symmetries form a group $G = \{I, S\}$ under composition of functions. (The complex numbers together with its two operations of addition and multiplication is a mathematical structure called a "field." The invertible functions I and S are *field isomorphisms*, meaning they preserve the two field operations on $\mathbb{C}$. See Chapter 6 for more on fields.) Indeed, the symmetries of a polynomial equation $f(x) = 0$ are always permutations (one-to-one and onto functions) of the set of zeroes of f, and the symmetries always form a finite group under composition of functions.

Galois showed that the zeroes of a polynomial equation $f(x) = 0$ can be expressed in terms of sums, differences, products, quotients and roots of the coefficients of f if and only if the symmetry group of f has a special property he called "solvability." Quadratic polynomials $f(x) = ax^2 + bx + c$ always have symmetry groups with this property—and the familiar quadratic formula expresses the zeroes of $f(x) = 0$ in terms of the coefficients a, b and c. Polynomials of degrees 3 and 4 also have "solvable" symmetry groups, and there do exist generalizations of the quadratic formula to these cases. However, there exist polynomials of degree 5 (and higher) whose symmetry groups are *not*solvable, so no such general formulas exist for polynomials of degree greater than 4.

Lie's goal was to attach a symmetry group to a differential equation in such a way that the group would contain information about the solutions to the equation. Unlike Galois' finite groups, Lie's groups were infinite *continuous* groups. Here's a simple example. Recall that we can interpret the differential equation $\mathbf{y}' = f(\mathbf{y})$ as a vector field for $\mathbf{y} \in \mathbb{R}^n$. A *symmetry* of the differential equation is a transformation $\mathbb{R}^n \to \mathbb{R}^n$ that leaves the vector field unchanged. Using the example from Howe's paper discussed in section 6, if $f(y_1, y_2) = (-y_2, y_1)$, then every rotation

$$\gamma(t) = \begin{bmatrix} \cos(t) & -\sin(t) \\ \sin(t) & \cos(t) \end{bmatrix}$$

for $t \in \mathbb{R}$ is a symmetry of $\mathbf{y}' = f(\mathbf{y})$. The rotations of the plane form a one-parameter subgroup. Quoting from *Transformation Groups for Beginners* by Duzhin and Chebotarevsky, it turns out that, in general, "knowledge of a one-parameter group of symmetries enables one to find the general solution of the equation under study in closed form."

According to Yaglom in his lovely book *Felix Klein and Sophus Lie: Evolution of the Idea of Symmetry in the Nineteenth Century* (pages 107–108)

> The beautiful Lie theory of differential equations was very highly valued by its creator. For a time it was extremely popular .... [but] already in the 1930s the main interest of mathematicians moved far away from Lie's constructions. ... But in the 1970s physicists and after them, of course, mathematicians, suddenly remembered "the Lie group of a differential equation."

What they realized was that the symmetries of the differential equation were also "symmetries of the real objects described (modeled) by the differential equation" leading to a "new surge of interest" in Lie's theory of differential equations. It's worth adding that although interest in Lie's theory of differential equations was in, out, and then in fashion, his theory itself, Lie groups and Lie algebras, has been a vital area of mathematics without interruption since its invention.

4.9 Appendix on convergence

Throughout what follows, fix a positive integer n. To use the definition

$$\exp(A) = I_n + A + \frac{1}{2!}A^2 + \cdots + \frac{1}{m!}A^m + \cdots$$

for A in $\mathcal{M}(n, \mathbb{R})$, we must give meaning to the *convergence* of a series of matrices. In order to do that, we need to be able to measure the size of the difference between two matrices — that is, to determine the *distance* between them. Since we're regarding $\mathcal{M}(n, \mathbb{R})$ as a subset of $\mathbb{R}^N$ for $N = n^2$, the most natural choice would be to use the Euclidean length $\|A\| = (\sum_{ij} a_{ij}^2)^{1/2}$. However, because of the square root, the Euclidean length is a bit messy to work with. Like some of the authors we have cited, we'll use a different definition (see especially Artin). It can be proved that a series converges using this definition of length if and only if it converges using the standard Euclidean length.

Definition of **length** for matrices. For $A = (a_{ij}) \in \mathcal{M}(n, \mathbb{R})$, let

$$\|A\| = n \max_{i,j} |a_{ij}|.$$

Notice that the definition implies $|a_{ij}| \leq (1/n)\|A\|$ for all i, j. (We're using double bars for the length of A because many books use $|A|$ to denote the determinant of A.)

Since our emphasis is on the algebra, we'll skim quite lightly through these convergence issues, but the crucial properties of the matrix length in the next proposition are pretty straightforward to verify.

Proposition 4.9.1 ***Properties of length.*** For all $A, B \in \mathcal{M}(n, \mathbb{R})$,

a. $\|A + B\| \leq \|A\| + \|B\|$ *(the **triangle inequality**).*

b. $\|AB\| \leq \|A\| \, \|B\|$.

c. $\|cA\| = |c|\|A\|$ *for all* $c \in \mathbb{R}$.

d. $\|A\| \geq 0$, *and* $\|A\| = 0$ *if and only if* A *is the zero matrix.*

Proof Property (a) follows from the triangle inequality for the absolute value of real numbers:

$$|a_{ij} + b_{ij}| \leq |a_{ij}| + |b_{ij}| \leq (1/n)(\|A\| + \|B\|)$$

so

$$\|A + B\| = n \max_{i,j} |a_{ij} + b_{ij}| \leq \|A\| + \|B\|.$$

The proof of (b) is a little notationally messy, but otherwise straightforward:

$$\begin{aligned}\|AB\| &= n \max_{i,j} \left|\sum_{k=1}^{n} a_{ik}b_{kj}\right| \\ &\leq n \max_{i,j} \sum_{k=1}^{n} |a_{ik}||b_{kj}| \\ &\leq n \sum_{k=1}^{n} \frac{\|A\|}{n} \cdot \frac{\|B\|}{n} \\ &\leq \|A\| \cdot \|B\|.\end{aligned}$$

The proof of (c) follows from the analogous property for absolute values of real numbers: $|ca_{ij}| = |c||a_{ij}|$.

The proof of (d) follows from the fact that $|a_{ij}| \geq 0$ and $|a_{ij}| = 0$ if and only if $a_{ij} = 0$. □

Using this notion of the *distance* $\|A - B\|$ between matrices A and B, we can talk about the *convergence* of a sequence (and hence a series) of matrices, which permits the definitions of $\exp(A)$ (and $\log(A)$). Our main result is the following.

Proposition 4.9.2 The series for $\exp(A)$ converges (absolutely) for every A in $\mathcal{M}(n, \mathbb{R})$.

Proof Write S_M for the partial sum

$$S_M = \sum_{k=0}^{M} \frac{1}{k!} A^k.$$

In order to show that the series for $\exp(A)$ converges, we need to show that the sequence $\{S_M\}$ of partial sums converges. We will show that $\{S_M\}$ is a **Cauchy sequence**: as M and N get larger and larger, the terms S_M and S_N of the sequence get closer and closer together. More precisely, given any (small) positive number ϵ there is some positive integer K so that if both M and N exceed K, then $||S_M - S_N|| < \epsilon$. It is a theorem of analysis that a sequence of real numbers converges if and only if it is a Cauchy sequence, and the corresponding theorem for a sequence of matrices in $\mathcal{M}(n, \mathbb{R})$ follows from this.

Assume that $M > N$, so that

$$S_M - S_N = \sum_{k=N+1}^{M} \frac{1}{k!} A^k.$$

Then, using Proposition 4.9.1, we have

$$||S_M - S_N|| \leq \sum_{k=N+1}^{M} \frac{1}{k!} ||A||^k.$$

Write $a = ||A|| \in \mathbb{R}$. We know from calculus that the series

$$e^a = \sum_{n=0}^{\infty} \frac{a^k}{k!}$$

converges, so the corresponding sequence of partial sums converges. Invoking the theorem of analysis referred to above, the sequence of partial sums for the series defining e^a is a Cauchy sequence. That means that given $\epsilon > 0$ we can choose a positive integer K so that $M, N > K$ implies

$$\left| \sum_{k=N+1}^{M} \frac{1}{k!} a^k \right| = \sum_{k=N+1}^{M} \frac{1}{k!} a^k = \sum_{k=N+1}^{M} \frac{1}{k!} ||A||^k < \epsilon.$$

This proves that $\{S_M\}$ is a Cauchy sequence, and so the series for $\exp(A)$ converges. To see that the series converges *absolutely*, just notice that the same argument works if we replace $A = (a_{ij})$ by the matrix of absolute values
$B = (|a_{ij}|)$, since $||A||$ is defined as n times the maximum of $|a_{ij}|$, so $\|A\| = \|B\|$. □

Proposition 4.9.2 implies that the series for $\exp(tA)$ also converges absolutely. We can therefore differentiate it term by term and establish $\exp(tA)' = A\exp(tA)$, which provides a formal argument for Proposition 4.2a.

Proposition 4.9.3 *The series*

$$log(A) = (A - I_n) - \frac{1}{2}(A - I_n)^2 + \frac{1}{3}(A - I_n)^3 - \cdots + \frac{(-1)^{m-1}}{m} A^m + \cdots$$

converges for every $A \in GL(n, \mathbb{R})$ *satisfying* $\|A - I_n\| < 1$. *(We write A rather than M for the matrix to avoid confusion with the positive integer we call M in the proof.)*

Proof We will proceed by showing that the sequence of partial sums for the series defining $\log(A)$ is a Cauchy sequence. As before, assume $M > N$, so

$$S_M - S_N = \sum_{k=N+1}^{M} \frac{(-1)^{k-1}}{k} (A - I_n)^k.$$

Using Proposition 4.9.1 we have

$$||S_M - S_N|| \leq \sum_{k=N+1}^{M} \frac{||A - I_n||^k}{k}.$$

Let $b = ||A - I_n|| \in \mathbb{R}$. We would like to relate the sum

$$\sum_{k=N}^{M} \frac{b^k}{k}$$

to a convergent series. Does $\sum_{k=1}^{\infty} b^k/k$ converge? Yes it does, provided that $|b| < 1$. Here's the proof. We know

$$\log(a) = \sum_{k=1}^{\infty} \frac{(-1)^{k-1}}{k} (a - 1)^k$$

converges for $a \in \mathbb{R}$ with $|a - 1| < 1$ (again, log means the *natural* logarithm). If we let $a - 1 = b$, then we have

$$\log(b + 1) = \sum_{k=1}^{\infty} \frac{(-1)^{k-1}}{k} b^k.$$

Replace b by $-b$ to get

$$\log(-b+1) = \sum_{k=1}^{\infty} \frac{(-1)^{k-1}}{k}(-1)^k b^k = \sum_{k=1}^{\infty} -\frac{b^k}{k}.$$

Therefore, we can conclude that $-\log(-b+1) = \sum_{k=1}^{\infty} b^k/k$ converges for $|b| < 1$. From this it follows that the sequence of partial sums for $\sum b^k/k$ is a Cauchy sequence, and thus the sequence of partial sums for $\log(A)$ is a Cauchy sequence as long as $|b| = ||A - I_n|| < 1$. □

There is a useful extra bit of information buried in the proof of Proposition 4.9.3. Notice that

$$||\log(A)|| \leq \sum_{k=1}^{\infty} \frac{||A - I_n||^k}{k} = -\log(-||A - I_n|| + 1),$$

so $||A - I_n|| < \frac{1}{2}$ implies $||\log(A)|| < \log(2)$. This fact is used in the proof of Lemma 4.9.4 below.

We now return to part (b) of Theorem 4.3, the general case.

Theorem 4.3b *If n is a positive integer and $\gamma: \mathbb{R} \to GL(n, \mathbb{R})$ is a one-parameter subgroup, then $\gamma(t) = \exp(tA)$, where $A = \gamma'(0)$.*

The following proof of the general case of Theorem 4.3 is from Howe's *Monthly* article "Very Basic Lie Theory." It requires two lemmas. Lemma 4.9.5 is rather technical, but Lemma 4.9.4 is quite important (and probably deserves a more dignified name than "lemma"). It gives the local relationship between a Lie group and its tangent space. This local correspondence is between a neighborhood of the zero matrix $0_n \in \mathcal{M}(n, \mathbb{R})$ and a neighborhood of the identity matrix $I_n \in GL(n, \mathbb{R})$. We need to make the word "neighborhood" precise. To do that we first define the *open ball* $\mathcal{B}_r(A)$ of radius $r > 0$ about a matrix $A \in \mathcal{M}(n, \mathbb{R})$:

$$\mathcal{B}_r(A) = \{B \in \mathcal{M}(n, \mathbb{R}) : \|B - A\| < r\}.$$

We also need the related idea of an open set of matrices: a set is *open* if it contains an open ball around each of its elements. Now we can state the two lemmas.

Lemma 4.9.4 *We can choose r sufficiently small so that exp maps $\mathcal{B}_r(0_n)$ one-to-one and onto an open set U' containing I_n.*

Lemma 4.9.5 *For $r < \log(2)$, matrices in $U' = \exp(\mathcal{B}_r(0_n))$ have* **unique** *square roots in U'.*

We omit the proofs of the lemmas (they are given in Howe's paper) and go on to give a sketch of the proof of Theorem 4.3b. You'll see that this sketch closely parallels the proof of Theorem 4.3a, the special case when $n = 1$.

Step 1. By the continuity of γ, for small h, $\gamma(h)$ is close to I_n. Using Lemma 4.9.4, if h is sufficiently small, we can find a matrix B near 0_n so that $\gamma(h) = \exp(B)$. If we choose $A = (1/h)B$, then $\gamma(h) = \exp(hA)$.

Step 2. Because γ is a homomorphism, we can use induction to show

$$\gamma(mh) = \exp(mhA)$$

for every positive integer m, and then we can extend the result to negative integers.

Step 3. Using Lemma 4.9.5 repeatedly, we can show that

$$\gamma(\frac{h}{2^k}) = \exp(\frac{h}{2^k}A)$$

for every positive integer k.

Step 4. Putting steps (2) and (3) together, we can show

$$\gamma(\frac{mh}{2^k}) = \exp(\frac{mh}{2^k}A)$$

for every integer m and every positive integer k.

Step 5. It can be shown that every real number is the limit of a sequence of rational numbers of the form $\frac{mh}{2^k}$. Using this fact and the continuity of γ, we can conclude that $\gamma(t) = \exp(tA)$ for every real number t. Because the series for $\exp(tA)$ converges absolutely, we can differentiate it term by term, and $\gamma'(t) = A\exp(tA)$ and $A = \gamma'(0)$. □

CHAPTER 5

Lie algebras

5.1 Lie algebras

The tangent space at the identity $L(G)$ captures some information about a Lie group G, but not enough to work backwards from the tangent space to the group. For example, the groups $SO(n, \mathbb{R})$ and $O(n, \mathbb{R})$ have the same tangent space. However, this coincidence reflects the fact that $O(n, \mathbb{R})$ falls into two pieces: the matrices of determinant $+1$ and the matrices of determinant -1. The tangent space of $O(n, \mathbb{R})$ is determined by the piece containing the identity, namely $SO(n, \mathbb{R})$, so their common tangent space isn't surprising. Other coincidences occur in pairs of groups whose tangent spaces we wouldn't expect to be the same, as in the following problem.

Problem 5.1.1.

a. What is the most significant difference you notice between $G_1 = SL(2, \mathbb{R})$ and

$$G_2 = \left\{ \begin{bmatrix} e^a & 0 & 0 \\ 0 & e^b & 0 \\ 0 & 0 & e^c \end{bmatrix} : a, b, c \in \mathbb{R} \right\}$$

as *groups*? (Assume that G_2 is a group under matrix multiplication.)

b. What are the tangent spaces $L(G_1)$ and $L(G_2)$? What are their dimensions?

c. Why are $L(G_1)$ and $L(G_2)$ isomorphic as vector spaces?

The fact that quite different groups can have tangent spaces that are isomorphic as vector spaces, as in (5.1.1), suggests that we should enrich the structure of the tangent space somehow. We do this by adding another operation to the two we already have (vector addition and scalar multiplication). The new operation associates to a pair of vectors another vector, resembling a vector product.

If L is a vector space over the real numbers and $\boldsymbol{v}$ and $\boldsymbol{w}$ are vectors in L, we write $[\boldsymbol{v}, \boldsymbol{w}] \in L$ for the **bracket** of $\boldsymbol{v}$ and $\boldsymbol{w}$. (Here's a case where the name of the mathematical idea is just the name of the notation.) We say the real vector space L is a **Lie algebra** if it has a bracket operation with the following properties.

- The vector space L is closed under the bracket:

$$\boldsymbol{v}, \boldsymbol{w} \in L \text{ implies } [\boldsymbol{v}, \boldsymbol{w}] \in L.$$

- The bracket operation is **anti-symmetric**:

$$[\boldsymbol{w}, \boldsymbol{v}] = -[\boldsymbol{v}, \boldsymbol{w}] \text{ for all } \boldsymbol{v}, \boldsymbol{w} \in L.$$

(Notice that this implies $[\boldsymbol{v}, \boldsymbol{v}] = \mathbf{0}$ for all $\boldsymbol{v} \in L$, where $\mathbf{0}$ is the zero vector in L. Do you see why?)

- The bracket operation is **bilinear**:

$$\begin{aligned}[\boldsymbol{u}, \boldsymbol{v} + \boldsymbol{w}] &= [\boldsymbol{u}, \boldsymbol{v}] + [\boldsymbol{u}, \boldsymbol{w}],\\ [\boldsymbol{u} + \boldsymbol{v}, \boldsymbol{w}] &= [\boldsymbol{u}, \boldsymbol{w}] + [\boldsymbol{v}, \boldsymbol{w}],\\ [r\boldsymbol{v}, \boldsymbol{w}] &= r[\boldsymbol{v}, \boldsymbol{w}] = [\boldsymbol{v}, r\boldsymbol{w}].\end{aligned}$$

for all $\boldsymbol{u}, \boldsymbol{v}, \boldsymbol{w} \in L$ and all $r \in \mathbb{R}$. (Bilinearity implies $[\boldsymbol{v}, \mathbf{0}] = \mathbf{0}$ for all $\boldsymbol{v} \in L$. Do you see why?)

- The bracket satisfies the **Jacobi identity**:

$$[\boldsymbol{u}, [\boldsymbol{v}, \boldsymbol{w}]] + [\boldsymbol{v}, [\boldsymbol{w}, \boldsymbol{u}]] + [\boldsymbol{w}, [\boldsymbol{u}, \boldsymbol{v}]] = \mathbf{0}$$

for all $\boldsymbol{u}, \boldsymbol{v}, \boldsymbol{w} \in L$.

Sometimes the bracket operation on a Lie algebra is called the **Lie bracket**.

For any choice of the positive integer n, when G is a subgroup of $GL(n, \mathbb{R})$, we have seen that $L(G)$ is a subspace of the vector space $\mathcal{M}(n, \mathbb{R})$. We begin by defining the **matrix bracket** $[A, B]$ of two matrices A and B in $\mathcal{M}(n, \mathbb{R})$:

$$[A, B] = AB - BA.$$

In the next section, we'll see where this definition comes from.

Problem 5.1.2. (S)

Show that the matrix bracket $[A, B] = AB - BA$ makes $\mathcal{M}(n, \mathbb{R})$ a Lie algebra by verifying the properties in (a)–(d).

a. The vector space $\mathcal{M}(n, \mathbb{R})$ is closed under the matrix bracket.

b. The matrix bracket is anti-symmetric.

c. The matrix bracket is bilinear.

d. The matrix bracket satisfies the Jacobi identity.

Notice that the matrix bracket $[A, B] = 0_n$ precisely when $AB = BA$. In fact, we could say that the matrix bracket $[A, B]$ measures how far A and B are from commuting with each other. This link to commuting matrices is why a Lie algebra L is called **abelian** if $[\boldsymbol{v}, \boldsymbol{w}] = \mathbf{0}$ for all $\boldsymbol{v}, \boldsymbol{w} \in L$. This language is used even when the Lie algebra doesn't consist of matrices.

Problem 5.1.3.

Show that a one-dimensional Lie algebra must be abelian.

Problem 5.1.4.

In this problem, you'll examine the relationship between the Jacobi identity and the associative law.

a. Give a *specific* example for $n = 2$ to show that the matrix bracket is *not* associative. That is, find matrices $A, B, C \in \mathcal{M}(2, \mathbb{R})$ for which $[A, [B, C]] \neq [[A, B], C]$.

b. **(C)** Can an anti-symmetric operation that satisfies the associative law also satisfy the Jacobi identity? In other words, assume that a vector space V is closed under an operation $\diamond$ that satisfies

$$\boldsymbol{w} \diamond \boldsymbol{v} = -(\boldsymbol{v} \diamond \boldsymbol{w}) \text{ for all } \boldsymbol{v}, \boldsymbol{w} \in V,$$
$$\boldsymbol{u} \diamond (\boldsymbol{v} \diamond \boldsymbol{w}) = (\boldsymbol{u} \diamond \boldsymbol{v}) \diamond \boldsymbol{w} \text{ for all } \boldsymbol{u}, \boldsymbol{v}, \boldsymbol{w} \in V.$$

Simplify $\boldsymbol{a} = \boldsymbol{u} \diamond (\boldsymbol{v} \diamond \boldsymbol{w}) + \boldsymbol{v} \diamond (\boldsymbol{w} \diamond \boldsymbol{u}) + \boldsymbol{w} \diamond (\boldsymbol{u} \diamond \boldsymbol{v})$ using these assumptions. Under what conditions would $\boldsymbol{a} = \boldsymbol{0}$ (the zero vector in V) for all $\boldsymbol{u}, \boldsymbol{v}, \boldsymbol{w} \in V$?

We shall see that the matrix bracket makes $L(G)$ a Lie algebra for each of the matrix groups G that we have studied.

Problem 5.1.5.(S)

In this problem you will show that the matrix bracket makes the tangent space of $G = \mathcal{L}(2, \mathbb{R})$ a Lie algebra. By (3.1.13),

$$L(G) = \left\{ \begin{bmatrix} 0 & b \\ b & 0 \end{bmatrix} : b \in \mathbb{R} \right\}.$$

We already know that $L(G)$ is a vector space and, from (5.1.2), that the matrix bracket has the properties (b)–(d). So all this problem requires is to show that $L(G)$ is *closed* under the bracket operation: if $A, B \in L(G)$, then $[A, B] \in L(G)$.

Problem 5.1.6.

In this problem you will show that the matrix bracket makes the tangent space of $G = SL(2, \mathbb{R})$ a Lie algebra. By (3.1.11), $L(G)$ consists of the 2 by 2 trace zero matrices.

$$L(G) = \left\{ \begin{bmatrix} a & b \\ c & -a \end{bmatrix} : a, b, c \in \mathbb{R} \right\}.$$

Show that $L(G)$ is closed under the matrix bracket.

Problem 5.1.7. (S)

Fix a positive integer n. Generalize (5.1.6) to $G = SL(n, \mathbb{R})$ for $n \geq 2$ by showing that

$$L(G) = \{B \in \mathcal{M}(n, \mathbb{R}) \,:\, \operatorname{tr}(B) = 0\}$$

is closed under the matrix bracket.

Problem 5.1.8.

Fix a positive integer n. Let $G = O(n, \mathbb{R})$, the Euclidean case, so

$$L(G) = \{B \in \mathcal{M}(n, \mathbb{R}) : B^T + B = 0_n\}.$$

Show $L(G)$ is a Lie algebra by showing it is closed under the matrix bracket.

Problem 5.1.9.

Fix a positive integer n. Let $C \in GL(n, \mathbb{R})$, and let $G = \{M \in GL(n, \mathbb{R}) : M^T CM = C\}$. By (4.5.7),

$$L(G) = \{B \in \mathcal{M}(n, \mathbb{R}) : B^T C + CB = 0_n\}.$$

Show $L(G)$ is a Lie algebra by showing it is closed under the matrix bracket.

The next problem introduces the real *Heisenberg group*

$$\mathcal{H} = \left\{ \begin{bmatrix} 1 & a & b \\ 0 & 1 & c \\ 0 & 0 & 1 \end{bmatrix} : a, b, c \in \mathbb{R} \right\}.$$

As you might expect from the use of Heisenberg's name, this group is used in quantum mechanics. We'll have more to say about the Heisenberg group and its applications in Chapter 6—see the remark after (6.3.4).

Problem 5.1.10.

a. Show that the Heisenberg group $\mathcal{H}$ defined above really is a group under matrix multiplication.

b. **(S)** Find the tangent space at the identity of the Heisenberg group.

$$\mathcal{H} = \left\{ \begin{bmatrix} 1 & a & b \\ 0 & 1 & c \\ 0 & 0 & 1 \end{bmatrix} : a, b, c \in \mathbb{R} \right\}.$$

c. Find a basis for $L(\mathcal{H})$.

d. Find the matrix bracket of each pair of basis vectors.

e. Is $L(\mathcal{H})$ closed under the matrix bracket? Why or why not?

The matrix bracket isn't the only operation that turns a vector space into a Lie algebra. Like the man who didn't realize he had been speaking prose all his life, you probably didn't realize that the familiar vector space $\mathbb{R}^3$ can be regarded as a Lie algebra by using the vector cross product as the bracket operation,

$$[\boldsymbol{v}, \boldsymbol{w}] = \boldsymbol{v} \times \boldsymbol{w}$$

for $\boldsymbol{v}, \boldsymbol{w} \in \mathbb{R}^3$.

Problem 5.1.11.

Let $L = \mathbb{R}^3$. Show that L is a Lie algebra via the bracket operation given by the vector cross product by verifying that it is anti-symmetric and bilinear and satisfies the Jacobi identity. Assume facts about the vector cross product from linear algebra, including the useful equation

$$\boldsymbol{u} \times (\boldsymbol{v} \times \boldsymbol{w}) = (\boldsymbol{u} \cdot \boldsymbol{w})\boldsymbol{v} - (\boldsymbol{u} \cdot \boldsymbol{v})\boldsymbol{w}.$$

With a new mathematical structure, the Lie algebra, comes a new kind of homomorphism, the Lie algebra homomorphism. Since a Lie algebra is a vector space with an additional operation, the bracket, it's not surprising that a Lie algebra homomorphism is a vector space homomorphism (linear transformation) that also preserves the bracket operation. More formally, suppose L_1 and L_2 are Lie algebras. A function $T : L_1 \to L_2$ is a **Lie algebra homomorphism** if T is a linear transformation and also T preserves the bracket:

$$T([\boldsymbol{v}, \boldsymbol{w}]) = [T(\boldsymbol{v}), T(\boldsymbol{w})]$$

for all $\boldsymbol{v}, \boldsymbol{w} \in L_1$, where the bracket on the left is the bracket operation in L_1, and the bracket on the right is the bracket operation in L_2. A **Lie algebra isomorphism** is a Lie algebra homomorphism that is one-to-one and onto.

Problem 5.1.12.

Let G be the Euclidean orthogonal group $O(3, \mathbb{R})$, so $L(G)$ consists of all the 3 by 3 anti-symmetric matrices: $A^T = -A$.

a. Explain why

$$L(G) = \left\{ \begin{bmatrix} 0 & -x & -y \\ x & 0 & -z \\ y & z & 0 \end{bmatrix} : x, y, z \in \mathbb{R} \right\}.$$

b. Explain why $L(G)$ and $\mathbb{R}^3$ are isomorphic as vector spaces.

c. **(S)** It's reasonable to wonder whether $L(G)$ and $\mathbb{R}^3$ are isomorphic as Lie algebras (where $\mathbb{R}^3$ is a Lie algebra via $[\boldsymbol{v}, \boldsymbol{w}] = \boldsymbol{v} \times \boldsymbol{w}$ by (5.1.11)). Indeed they are. Define $T : \mathbb{R}^3 \to L(G)$ by

$$T(\boldsymbol{v}) = T(x, y, z) = \begin{bmatrix} 0 & -x & -y \\ x & 0 & -z \\ y & z & 0 \end{bmatrix}$$

where $\boldsymbol{v} = (x, y, z) \in \mathbb{R}^3$. Show T is a Lie algebra isomorphism.

Problem 5.1.13.

a. Explain why $L(\mathcal{H})$ (see (5.1.10)) and $\mathbb{R}^3$ are isomorphic as vector spaces.

b. **(C)** Are $\mathbb{R}^3$ (viewed as a Lie algebra as in (5.1.11)) and $L(\mathcal{H})$ isomorphic as Lie algebras? Why or why not?

The vector spaces we have been working with are finite-dimensional. The following problem shows that an infinite-dimensional vector space can be quite different from a finite-dimensional one.

Problem 5.1.14.

(This problem is adapted from MIT Open Courseware, the Linear Algebra page of Dr. Anna Lachowska.) Fix a positive integer n.

a. Fix an arbitrary positive integer n. Explain why it is not possible to find two n by n matrices A and B with $[A, B] = I_n$. (Equivalently, it is not possible to find two linear transformations $S, T : \mathbb{R}^n \to \mathbb{R}^n$ with $[S, T] = S \circ T - T \circ S = I$, where $I(\boldsymbol{v}) = \boldsymbol{v}$ for all $\boldsymbol{v} \in \mathbb{R}^n$.)

b. The set of polynomials

$$\mathcal{P} = \{p(x) = a_m x^m + \cdots + a_1 x + a_0 \ : \ m \in \mathbb{Z}, m \geq 0, a_j \in \mathbb{R}\},$$

with addition and scalar multiplication of polynomials defined in the usual way, is a real vector space. Explain why $\mathcal{P}$ is infinite-dimensional.

c. Define functions $S, T : \mathcal{P} \to \mathcal{P}$ by $S(p(x)) = xp(x) = a_m x^{m+1} + \cdots + a_1 x^2 + a_0 x$ and $T(p(x)) = p'(x) = m a_m x^{m-1} + \cdots + a_1$. Show that S and T are linear transformations. (You can assume facts from calculus.)

d. Continuing the notation of (c), show $[S, T] = I$. (So, if we imagine "infinite" matrices representing S and T, we have a different result than in (a) for finite matrices.)

5.2 Adjoint maps—big 'A' and small 'a'

In this section we will see where the matrix bracket operation comes from, but the path is a bit roundabout. We start by going back to linear algebra, where you may have noticed the close relationship between an n by n matrix X and its *conjugate* MXM^{-1} via a matrix $M \in GL(n, \mathbb{R})$:

- If X is regarded as the matrix of a linear transformation $\mathbb{R}^n \to \mathbb{R}^n$ with respect to some basis of $\mathbb{R}^n$, and Y is another matrix for that transformation (with respect to a different basis of $\mathbb{R}^n$), then $Y = MXM^{-1}$ for some invertible matrix M; i.e., X and Y are conjugate.
- A matrix and its conjugate have the same determinant: $\det(X) = \det(MXM^{-1})$.
- A matrix and its conjugate have the same trace: $\operatorname{tr}(X) = \operatorname{tr}(MXM^{-1})$.

This list just scratches the surface of the relationship between a matrix and its conjugates. (If you've studied abstract algebra, you've encountered "conjugacy classes" in groups and the related notion of a *normal subgroup*: a subgroup containing the conjugates of each of its elements.) We will formalize the notion of conjugate in our setting by making the following definition.

Fix a positive integer n and a matrix $M \in GL(n, \mathbb{R})$. Define a function

$$\operatorname{Ad}(M) : \mathcal{M}(n, \mathbb{R}) \to \mathcal{M}(n, \mathbb{R})$$

by

$$\operatorname{Ad}(M)(X) = MXM^{-1}$$

for $X \in \mathcal{M}(n, \mathbb{R})$. The mapping Ad taking the matrix M to the function $\operatorname{Ad}(M)$ is called the **Adjoint** map. The capital 'A' is important because there is also a small 'a' adjoint map, as we'll see later.

Problem 5.2.1. (S)
Fix a positive integer n and a matrix $M \in GL(n, \mathbb{R})$. In parts (a) and (b) of this problem, you'll show that $\operatorname{Ad}(M) : \mathcal{M}(n, \mathbb{R}) \to \mathcal{M}(n, \mathbb{R})$ is a Lie algebra homomorphism—that is, $\operatorname{Ad}(M)$ is a linear transformation and preserves the matrix bracket. In part (c) you'll show it is invertible, and hence one-to-one and onto, so it is a Lie algebra isomorphism.

a. Show that $\mathrm{Ad}(M)$ is a linear transformation.

b. Show that $\mathrm{Ad}(M)$ preserves the matrix bracket on $\mathcal{M}(n, \mathbb{R})$.

c. Show that $\mathrm{Ad}(M)$ is invertible. (Hint: What function would "un-do" $\mathrm{Ad}(M)$?)

As a vector space, $\mathcal{M}(n, \mathbb{R})$ is essentially $\mathbb{R}^{n^2}$, so parts (a) and (c) of (5.2.1) show that

$$\mathrm{Ad} : GL(n, \mathbb{R}) \to GL(\mathbb{R}^{n^2}).$$

Recall that the elements of $GL(n, \mathbb{R})$ are matrices, so the binary operation in the group $GL(n, \mathbb{R})$ is matrix multiplication, and the elements of the group $GL(\mathbb{R}^{n^2})$ are linear transformations, so the binary operation in the group $GL(\mathbb{R}^{n^2})$ is composition of functions.

Problem 5.2.2. (S)

Fix a positive integer n. Show that the function Ad: $GL(n, \mathbb{R}) \to GL(\mathbb{R}^{n^2})$ is a group homomorphism. That is, show that $\mathrm{Ad}(M_1 M_2) = \mathrm{Ad}(M_1) \circ \mathrm{Ad}(M_2)$ for all $M_1, M_2 \in GL(n, \mathbb{R})$. (Hint: You're to verify the equality of two functions, so you need to show that both functions have the same effect on $X \in \mathcal{M}(n, \mathbb{R})$. Remember that $M_1 M_2$ is the result of matrix multiplication, while $\mathrm{Ad}(M_1)\circ \mathrm{Ad}(M_2)$ is the result of composition of functions.)

Now we come to the promised **small 'a' adjoint** map denoted ad. We begin with a **provisional definition** that is easy to work with. Suppose L is a Lie algebra. For a fixed $\boldsymbol{v} \in L$ define $\mathrm{ad}(\boldsymbol{v}) : L \to L$ by

$$\mathrm{ad}(\boldsymbol{v})(\boldsymbol{w}) = [\boldsymbol{v}, \boldsymbol{w}]$$

for all $\boldsymbol{w} \in L$. The adjoint map is thus intimately tied to the Lie algebra structure given by the bracket operation.

Problem 5.2.3.

Suppose L is a Lie algebra and fix $\boldsymbol{v} \in L$. Use the properties of the bracket to show $\mathrm{ad}(\boldsymbol{v})$ is a linear transformation $L \to L$.

Most of our examples are Lie algebras of matrices, subspaces of $\mathcal{M}(n, \mathbb{R})$ for some positive integer n. Applying the definition in this situation, for a fixed matrix $A \in \mathcal{M}(n, \mathbb{R})$,

$$\mathrm{ad}(A) : \mathcal{M}(n, \mathbb{R}) \to \mathcal{M}(n, \mathbb{R}) \text{ by } \mathrm{ad}(A)(X) = [A, X] = AX - XA$$

for all $X \in \mathcal{M}(n, \mathbb{R})$. We know from (5.2.3) that $\mathrm{ad}(A)$ is a linear transformation. Because $\mathcal{M}(n, \mathbb{R})$ is essentially $\mathbb{R}^{n^2}$, (5.2.3) shows that

$$\mathrm{ad} : \mathcal{M}(n, \mathbb{R}) \to \mathcal{M}(n^2, \mathbb{R}),$$

where we interpret the elements of $\mathcal{M}(n^2, \mathbb{R})$ as linear transformations $\mathbb{R}^{n^2} \to \mathbb{R}^{n^2}$. (These linear transformations needn't be invertible, so we're not talking about the group $GL(\mathbb{R}^{n^2})$ here.)

Problem 5.2.4.
Fix a positive integer n. Use the formula $\mathrm{ad}(A)(X) = [A, X]$ for all $X \in \mathcal{M}(n, \mathbb{R})$ to show that the function ad preserves the Lie bracket. That is, show $\mathrm{ad}([A, B]) = [\mathrm{ad}(A), \mathrm{ad}(B)]$ for any $A, B \in \mathcal{M}(n, \mathbb{R})$. Because $\mathrm{ad}(A)$ and $\mathrm{ad}(B)$ are functions, interpret the bracket on the right as $\mathrm{ad}(A) \circ \mathrm{ad}(B) - \mathrm{ad}(B) \circ \mathrm{ad}(A)$.

Problem 5.2.5.
Fix a positive integer m. Suppose L is a Lie algebra and, assume the dimension of L as a vector space is m. Suppose $\boldsymbol{v}_1, \ldots, \boldsymbol{v}_m$ is a basis for L. Explain why knowing how to calculate the bracket of any two basis vectors (i.e., knowing $[\boldsymbol{v}_i, \boldsymbol{v}_j]$ for all $i, j = 1, \ldots, m$) enables you to calculate the bracket $[\boldsymbol{u}, \boldsymbol{w}]$ of *any* two elements $\boldsymbol{u}$ and $\boldsymbol{w}$ of L.

Problem 5.2.6.
Fix a positive integer m. Show that if two abelian Lie algebras have the same vector space dimension m over the real numbers, then they are isomorphic as Lie algebras.

Problem 5.2.7.
In this problem you will take a careful look at the Lie algebra of $G = SL(2, \mathbb{R})$, using the following basis for $L(G) \subseteq \mathcal{M}(2, \mathbb{R})$:

$$E = \begin{bmatrix} 0 & 1 \\ 0 & 0 \end{bmatrix}, \quad H = \begin{bmatrix} 1 & 0 \\ 0 & -1 \end{bmatrix}, \quad F = \begin{bmatrix} 0 & 0 \\ 1 & 0 \end{bmatrix}.$$

a. **(S)** Find the matrix brackets $[E, H]$, $[F, H]$ and $[E, F]$.
b. Assume $A \in L(G)$. Explain why $\mathrm{ad}(A) : L(G) \to L(G)$.
c. **(S)** Using (b) along with (5.2.3), we know $\mathrm{ad}(E)$ is a linear transformation of $L(G)$. Using the basis E, H, F (in that order), find the matrix of $\mathcal{E} = \mathrm{ad}(E)$. (The matrix will be 3 by 3 since $\dim L(G) = 3$.)
d. Find the matrix of $\mathcal{H} = \mathrm{ad}(H)$ and the matrix of $\mathcal{F} = \mathrm{ad}(F)$, as in (c).
e. **(S)** Find $\exp(\mathcal{E})$, $\exp(\mathcal{H})$ and $\exp(\mathcal{F})$.

Problem 5.2.8.
Continue the notation of (5.2.7), but now we bring Ad into the story, where for $M \in GL(2, \mathbb{R})$ and $X \in \mathcal{M}(2, \mathbb{R})$, $\mathrm{Ad}(M)(X) = MXM^{-1}$.
a. Explain why $\mathrm{Ad}(M) : L(G) \to L(G)$.
b. **(S)** Find $M_E = \exp(E)$, $M_H = \exp(H)$, and $M_F = \exp(F)$.
c. **(S)** Find the matrix of $\mathrm{Ad}(M_E)$ with respect to the ordered basis E, H, F of L(G).
d. Find the matrices of $\mathrm{Ad}(M_H)$ and $\mathrm{Ad}(M_F)$ with respect to the ordered basis E, H, F of L(G).

Pattern alert: Look back over your results in (5.2.7) and (5.2.8). Did you find $\mathrm{Ad}(\exp(U)) = \exp(\mathrm{ad}(U))$ for the special cases $U \in \{E, H, F\}$?

Finally, we are going to use the Adjoint and adjoint maps to motivate the definition of the matrix bracket. To do this, we back up and give a new **formal definition of the small 'a' adjoint**. In our new approach, we'll define ad as the differential of Ad. (This formal definition always agrees with the provisional one given earlier, although we'll only verify it in the case of Lie algebras of matrices.) First we set up the notation. Fix a positive integer n, and let $G_1 = GL(n, \mathbb{R})$ and $G_2 = GL(\mathbb{R}^{n^2})$. Start with

$$\mathrm{Ad} : G_1 \to G_2$$

as in (5.2.2). We know Ad is a group homomorphism. *Assume* Ad is a Lie group homomorphism, and **define the function ad** as the differential of Ad,

$$\mathrm{ad} : L(G_1) \to L(G_2).$$

(Writing ad=d(Ad) looks a bit funny.) We need to figure out what ad does to an element of $L(G_1)$. Choose $A \in L(G_1) = \mathcal{M}(n, \mathbb{R})$. We know that $\gamma_1(t) = \exp(tA)$ is a smooth curve in $G_1 = GL(n, \mathbb{R})$ passing through the identity I_n with $\gamma_1'(0) = A$. Let $\gamma_2(t) = \mathrm{Ad}(\gamma_1(t))$. By the definition of the differential, $\mathrm{ad}(A) = \gamma_2'(0)$.

Problem 5.2.9.
Continuing the notation above, for $B \in \mathcal{M}(n, \mathbb{R})$,

$$\gamma_2(t)(B) = \mathrm{Ad}(\gamma_1(t))(B).$$

a. Rewrite the right-hand side of this equation, then differentiate with respect to t, and finally evaluate at $t = 0$. What does this give you for $\gamma_2'(0)(B)$?
b. Explain why the calculation in (a) shows that $\mathrm{ad}(A)(B) = AB - BA$.

The result of (5.2.9) tells us that the two definitions of the adjoint agree, provided we define the matrix bracket by $[A, B] = AB - BA$.

Problem 5.2.10.
Fix a positive integer n. Let G be a subgroup of $GL(n, \mathbb{R})$ and let $L(G)$ be the Lie algebra of G. For $X \in L(G)$, let $\mathcal{X} = (\mathcal{X}_{ij})$ be the matrix of the linear transformation $\mathrm{ad}(X) : L(G) \to L(G)$ with respect to some basis of $L(G)$. Assume that $\mathcal{X}$ is *strictly upper triangular*, that is that $\mathcal{X}_{ij} = 0$ for $i \geq j$. Show that $\det(\exp(\mathcal{X})) = 1$.

Problem 5.2.11. (C)
a. Assume Theorem 4.4*. Show that if a subgroup G of $GL(n, \mathbb{R})$ is an abelian group, then its Lie algebra $L(G)$ is an abelian Lie algebra. (Hint: For $A, B \in L(G)$ with $A = \alpha'(0)$ and $B = \beta'(0)$, define $\gamma(t) = \alpha(t)\beta(t)\alpha(-t)\beta(-t)$ and calculate first $\gamma''(t)$ and then $\gamma''(0)$. This isn't a typo; those really are second derivatives.)
b. Give an explicit counterexample to the converse of (a). (See Theorem 5.11* in section 4 for the correct statement of a partial converse.)

Problem 5.2.12. (C)
Both the Heisenberg group of (5.1.9) and $SL(2, \mathbb{R})$ have Lie algebras of dimension 3. Are $L(\mathcal{H})$ and the Lie algebra of $SL(2, \mathbb{R})$ isomorphic as Lie algebras? Why or why not?

As noted in (5.2.12), the Lie algebra of $SL(2, \mathbb{R})$ is a 3-dimensional vector space. Obviously $\mathbb{R}^3$ is a 3-dimensional vector space, so we know from linear algebra that $\mathbb{R}^3$

and the Lie algebra of $SL(2, \mathbb{R})$ are isomorphic as vector spaces. If we regard $\mathbb{R}^3$ as a Lie algebra as in (5.1.10), it's natural to wonder whether these two Lie algebras are isomorphic as Lie algebras. In fact they *are*, but somewhat surprisingly the proof requires using complex numbers to describe $\mathbb{R}^3$. The argument is in the next problem.

Problem 5.2.13.
Choose the standard basis $\boldsymbol{v}_1, \boldsymbol{v}_2, \boldsymbol{v}_3$ for $\mathbb{R}^3$, so $[\boldsymbol{v}_1, \boldsymbol{v}_2] = \boldsymbol{v}_3$, $[\boldsymbol{v}_2, \boldsymbol{v}_3] = \boldsymbol{v}_1$, and $[\boldsymbol{v}_3, \boldsymbol{v}_1] = \boldsymbol{v}_2$. Define $\boldsymbol{u}_1 = i\boldsymbol{v}_2 - \boldsymbol{v}_3$, $\boldsymbol{u}_2 = 2i\boldsymbol{v}_1$, $\boldsymbol{u}_3 = i\boldsymbol{v}_2 + \boldsymbol{v}_3$. Assume $[i\boldsymbol{v}, \boldsymbol{w}] = i[\boldsymbol{v}, \boldsymbol{w}] = [\boldsymbol{v}, i\boldsymbol{w}]$ for all $\boldsymbol{v}, \boldsymbol{w} \in \mathbb{R}^3$.

a. Calculate $[\boldsymbol{u}_2, \boldsymbol{u}_1]$, $[\boldsymbol{u}_1, \boldsymbol{u}_3]$ and $[\boldsymbol{u}_2, \boldsymbol{u}_3]$.

b. Let $V = \{\sum_j a_j \boldsymbol{u}_j : a_j \in \mathbb{R}\}$. You can assume that V is a vector space over $\mathbb{R}$ with basis $\boldsymbol{u}_1$, $\boldsymbol{u}_2$, $\boldsymbol{u}_3$ and a Lie algebra under the bracket as in (a). Define $T: L(SL(2, \mathbb{R})) \to V$ by $T(E) = \boldsymbol{u}_1$, $T(H) = \boldsymbol{u}_2$, and $T(F) = \boldsymbol{u}_3$, with notation as in (5.2.7). Show that T is a Lie algebra isomorphism.

c. You may be wondering where the vectors $\boldsymbol{u}_j$ came from. To begin to see how the $\boldsymbol{u}_j$ were determined, find the matrix A of $\mathrm{ad}(\boldsymbol{v}_1)$ with respect to the basis $\boldsymbol{v}_1$, $\boldsymbol{v}_2$, $\boldsymbol{v}_3$, and then find the eigenvalues and eigenvectors of A. (Similar ideas occur in Example 6.5.3 in section 6.5.)

5.3 Putting the pieces together

Definition We say the (real) vector space L is a **Lie algebra** if it has a bracket operation with the the following properties.

- The vector space L is closed under the bracket: $\boldsymbol{v}, \boldsymbol{w} \in L$ implies $[\boldsymbol{v}, \boldsymbol{w}] \in L$.
- The bracket operation is **anti-symmetric**: $[\boldsymbol{w}, \boldsymbol{v}] = -[\boldsymbol{v}, \boldsymbol{w}]$ for all $\boldsymbol{v}, \boldsymbol{w} \in L$. (This implies $[\boldsymbol{v}, \boldsymbol{v}] = \boldsymbol{0}$ for all $\boldsymbol{v} \in L$.)
- The bracket operation is **bilinear**:

$$\begin{aligned}
[\boldsymbol{u}, \boldsymbol{v} + \boldsymbol{w}] &= [\boldsymbol{u}, \boldsymbol{v} + [\boldsymbol{u}, \boldsymbol{w}], \\
[\boldsymbol{u} + \boldsymbol{v}, \boldsymbol{w}] &= [\boldsymbol{u}, \boldsymbol{w}] + [\boldsymbol{v}, \boldsymbol{w}], \\
[r\boldsymbol{v}, \boldsymbol{w}] &= r[\boldsymbol{v}, \boldsymbol{w}] = [\boldsymbol{v}, r\boldsymbol{w}].
\end{aligned}$$

  for all $\boldsymbol{u}, \boldsymbol{v}, \boldsymbol{w} \in L$ and all $r \in \mathbb{R}$.
- The bracket satisfies the **Jacobi identity**:

$$[\boldsymbol{u}, [\boldsymbol{v}, \boldsymbol{w}]] + [\boldsymbol{v}, [\boldsymbol{w}, \boldsymbol{u}]] + [\boldsymbol{w}, [\boldsymbol{u}, \boldsymbol{v}]] = \boldsymbol{0}$$

  for all $\boldsymbol{u}, \boldsymbol{v}, \boldsymbol{w} \in L$ (where $\boldsymbol{0}$ is the zero vector in L).

A Lie algebra is called **abelian** if $[\boldsymbol{v}, \boldsymbol{w}] = \boldsymbol{0}$ for all $\boldsymbol{v}, \boldsymbol{w} \in L$. A subspace of a Lie algebra L that is closed under the bracket operation is called a **Lie subalgebra** of L.

Note that in general the bracket is *not* associative. An exception is when the Lie algebra is abelian, so all three of $[\boldsymbol{u}, [\boldsymbol{v}, \boldsymbol{w}]]$, $[\boldsymbol{v}, [\boldsymbol{w}, \boldsymbol{u}]]$ and $[\boldsymbol{w}, [\boldsymbol{u}, \boldsymbol{v}]]$ equal $\boldsymbol{0}$ for all $\boldsymbol{u}, \boldsymbol{v}, \boldsymbol{w} \in L$.

For any positive integer n, when G is a subgroup of $GL(n, \mathbb{R})$, we have seen that $L(G)$ is a subspace of the vector space $\mathcal{M}(n, \mathbb{R})$. In this case, the **matrix bracket** $[A, B]$ of two matrices A and B in $\mathcal{M}(n, \mathbb{R})$ is $[A, B] = AB - BA$.

Proposition 5.1 *Fix a positive integer n.*

a. *The vector space $\mathcal{M}(n, \mathbb{R})$ together with the bracket operation $[A, B] = AB - BA$ for $A, B \in \mathcal{M}(n, \mathbb{R})$ is a Lie algebra.*

b. *For G equal to any one of $SL(n, \mathbb{R})$, $\mathcal{L}(n, \mathbb{R})$ $(n = 2, 3, 4)$, $O(n, \mathbb{R})$ (Euclidean or not), $SO(n, \mathbb{R})$, or $Sp(n, \mathbb{R})$, the corresponding tangent space $L(G)$ is a Lie subalgebra of $\mathcal{M}(n, \mathbb{R})$.*

c. *The vector space $\mathbb{R}^3$ together with the bracket operation $[\boldsymbol{v}, \boldsymbol{w}] = \boldsymbol{v} \times \boldsymbol{w}$ is a Lie algebra.*

d. *Every one-dimensional Lie algebra is abelian.*

Although we are not able to prove it using our methods, the following more general theorem is true.

Theorem 5.2* *Fix a positive integer n. If G is any matrix Lie group, $G \subseteq GL(n, \mathbb{R})$, then $L(G)$ is a Lie subalgebra of $\mathcal{M}(n, \mathbb{R})$.*

For a matrix group G, we refer to $L(G)$ as the **Lie algebra of** G.

Definition—big 'A' Adjoint Fix a positive integer n. For $M \in GL(n, \mathbb{R})$, define a function

$$\mathrm{Ad}(M) : \mathcal{M}(n, \mathbb{R}) \to \mathcal{M}(n, \mathbb{R})$$

by

$$\mathrm{Ad}(M)(X) = MXM^{-1}$$

for $X \in \mathcal{M}(n, \mathbb{R})$. The function Ad taking M to $\mathrm{Ad}(M)$ is called the **Adjoint** map.

Proposition 5.3 *Fix a positive integer n. Ad(M) is an invertible linear transformation of the vector space $\mathcal{M}(n, \mathbb{R})$.*

Since the vector space $\mathcal{M}(n, \mathbb{R})$ is really the same as $\mathbb{R}^{n^2}$, Proposition 5.2 tells us that for $M \in GL(n, \mathbb{R})$, we have $\mathrm{Ad}(M) \in GL(\mathbb{R}^{n^2})$. In other words, Ad is the name of a function that maps a matrix M in $GL(n, \mathbb{R})$ to a linear transformation $\mathrm{Ad}(M)$ in $GL(\mathbb{R}^{n^2})$. We express this by writing

$$\mathrm{Ad} : GL(n, \mathbb{R}) \to GL(\mathbb{R}^{n^2}).$$

Proposition 5.4 *Fix a positive integer n. Ad: $GL(n, \mathbb{R}) \to GL(\mathbb{R}^{n^2})$ is a Lie group homomorphism.*

Definition—small 'a' adjoint The differential of the Adjoint map Ad is called the **adjoint** and is denoted ad. The differential takes $\mathcal{M}(n, \mathbb{R})$, the tangent space of $GL(n, \mathbb{R})$, to $\mathcal{M}(n^2, \mathbb{R})$, the tangent space of $GL(\mathbb{R}^{n^2}) \simeq GL(n^2, \mathbb{R})$:

$$\mathrm{ad} : \mathcal{M}(n, \mathbb{R}) \to \mathcal{M}(n^2, \mathbb{R}).$$

By Theorem 4.7*, the relationship between Ad, ad and exp is summarized by the following commutative diagram.

$$\begin{array}{ccc} GL(n,\mathbb{R}) & \overset{\text{Ad}}{\longrightarrow} & GL(\mathbb{R}^{n^2}) \simeq GL(n^2,\mathbb{R}) \\ \exp \uparrow & & \exp \uparrow \\ \mathcal{M}(n,\mathbb{R}) & \overset{\text{ad}}{\longrightarrow} & \mathcal{M}(n^2,\mathbb{R}) \end{array}$$

Proposition 5.5 *Fix a positive integer n. The following statements are true for all $A, B \in \mathcal{M}(n,\mathbb{R})$.*

a. ad *is a linear transformation.*

b.* $\text{Ad}\,(\exp(A)) = \exp(\text{ad}\,(A))$.

c. $\text{ad}\,(A)(B) = AB - BA = [A, B]$.

Remarks

- Part (b) of Proposition 5.5 has an asterisk, since it expresses the meaning of the commutative diagram above that follows from Theorem 4.7*.
- Part (c) of Proposition 5.5 motivates the definition of the Lie bracket for matrices in $\mathcal{M}(n,\mathbb{R})$.
- The matrix $\text{ad}(A)(B) = [A, B] = AB - BA$ is zero exactly when A and B commute. If A and B don't commute, then there is a sense in which $\text{ad}(A)(B) = [A, B]$ measures the "amount" by which A and B fail to commute.
- The definitions of the Adjoint and the adjoint can be generalized. We'll look at the adjoint map on more general Lie algebras in Chapter 6.

Definition If L_1 and L_2 are Lie algebras (over $\mathbb{R}$) and

$$T : L_1 \to L_2,$$

we say T is a **Lie algebra homomorphism** if T is a linear transformation and T also preserves the bracket operation:

$$T([\boldsymbol{v}, \boldsymbol{w}]) = [T(\boldsymbol{v}), T(\boldsymbol{w})]$$

for all $\boldsymbol{v}, \boldsymbol{w} \in L_1$. If T is also invertible, we say it is a **Lie algebra isomorphism**, and we say L_1 and L_2 are **isomorphic as Lie algebras**.

Proposition 5.6 *Fix a positive integer n. For every $M \in GL(n,\mathbb{R})$, $Ad(M)$ is a Lie algebra isomorphism.*

Proposition 5.7 *If L is a Lie algebra, the Lie bracket on L is determined by its effect on a basis of L. More specifically, assume the dimension of L is the positive integer m and $\boldsymbol{v}_1, \ldots, \boldsymbol{v}_m$ is a basis of L. If $B = \sum b_i \boldsymbol{v}_i$ and $C = \sum c_j \boldsymbol{v}_j$ are any two elements of L, then*

$$[B, C] = \sum_{i,j=1}^{m} b_i c_j [\boldsymbol{v}_i, \boldsymbol{v}_j].$$

Proposition 5.8 *Two abelian Lie algebras of the same dimension are isomorphic as Lie algebras.*

Proposition 5.9* (The proof depends on Theorem 4.4*.) *If* $G \subset GL(n, \mathbb{R})$ *is an abelian Lie group, then its Lie algebra* $L(G) \subseteq \mathcal{M}(n, \mathbb{R})$ *is abelian.*

The partial converse to Proposition 5.9* appears in the next section.

Suggestions for further reading

Humphreys, J.E., *Introduction to Lie Algebras and Representation Theory,* Graduate Texts in Mathematics **9**, Springer-Verlag (1972). Chapter 1 provides a terse but very clearly-written introduction to abstract Lie algebras.

Rossmann, W., *Lie Groups: An Introduction through Linear Groups*, Oxford University Press (2002). Chapter 1 on the exponential map includes a somewhat more general treatment of both the adjoint and the Adjoint. It also includes some of the analytic details and makes the link to vector fields.

5.4 A broader view: Lie theory

We are just beginning to glimpse the many ways in which the Lie algebra of a Lie group encodes significant information about the group itself. In this section we will take a quick tour of some of the important theorems on the relationship between Lie groups and their Lie algebras. Their hypotheses often require that the Lie group(s) in question be simply connected. (See section 3.5.) Many of the groups we have studied are connected, but most of them are *not* simply connected, so you might worry that these theorems are of limited applicability. However, Theorem 5.10* below tells us that every connected Lie group is closely related to a simply connected Lie group called its *universal cover*.

Theorem 5.10*

a. *If* L *is a Lie algebra, then there exists a simply connected Lie group* $G(L)$ *with Lie algebra isomorphic (as a Lie algebra) to* L*, and the group* $G(L)$ *is unique up to Lie group isomorphism.*
b. *If* G *is a connected Lie group with Lie algebra* $L = L(G)$*, then* G *is a homomorphic image of the simply connected Lie group* $G(L)$*, called the universal cover of* G*. The kernel of the homomorphism—the elements of* $G(L)$ *that map onto the identity in* G*—is a discrete subgroup of* $G(L)$*.*

The ideas in Theorem 5.10* can be illustrated by a simple example. The group $SO(2, \mathbb{R})$, which is isomorphic to the circle S^1, is connected but not simply connected. The familiar one-parameter subgroup $\gamma : \mathbb{R} \to SO(2, \mathbb{R})$ defined by

$$\gamma(t) = \begin{bmatrix} \cos(t) & -\sin(t) \\ \sin(t) & \cos(t) \end{bmatrix}$$

is a Lie group homomorphism from the additive group of real numbers, which *is* simply connected, onto $SO(2, \mathbb{R})$. The kernel of this homomorphism is

$$\Gamma = \{t \in \mathbb{R} : \gamma(t) = I_2\} = \{2n\pi : n \in \mathbb{Z}\}.$$

This kernel Γ is a group isomorphic to the integers under addition. (It is traditional to use uppercase gamma for the kernel of this homomorphism.) Both Γ and $\mathbb{Z}$ are *discrete* because there are "spaces" between every two group elements—think of the integers spread out on the real line. So the universal covering group of $SO(2, \mathbb{R})$ is the group $\mathbb{R}$. (If you've studied abstract algebra, you know this says that $SO(2, \mathbb{R}) \cong \mathbb{R}/\Gamma$.) The groups $SO(2, \mathbb{R})$ and $\mathbb{R}$ have isomorphic one-dimensional (abelian) Lie algebras.

A more complicated example is the connected, but not simply connected, group $SO(3, \mathbb{R})$. Its universal cover is the simply connected group $SU(2, \mathbb{C})$, which we know is isomorphic to S^3 by (2.2.11). The kernel of the homomorphism from $SU(2, \mathbb{C})$ onto $SO(3, \mathbb{R})$ is the discrete group $\{I_2, -I_2\}$. (By (3.2.13) there is a homomorphism from S^3 onto $SO(3, \mathbb{R})$ with kernel $\{1, -1\}$; in the isomorphism between $SU(2, \mathbb{C})$ and S^3, I_2 corresponds to 1 and $-I_2$ corresponds to -1.)

To illustrate how we can use these theorems to characterize connected groups by their Lie algebras, consider the following partial converse to Proposition 5.9*. Theorem 5.11* is starred because it depends on theorems we haven't proved, but the way the pieces fit together in the proof is accessible using our methods.

Theorem 5.11* *Assume the Lie group G is connected and its Lie algebra $L(G)$ is abelian of dimension $n > 0$. Then G is also abelian.*

Proof This argument doesn't require that G be a matrix group, so let's write $*$ for the operation in G. We know that the abelian group of n by n diagonal matrices with positive entries given in (4.5.8) has an abelian Lie algebra of dimension n. Further, this group of diagonal matrices (under matrix multiplication) is isomorphic to the group $\mathbb{R}^n$ (under addition), and $\mathbb{R}^n$ is simply connected. We also know from (5.2.11c) that any two abelian Lie algebras of the same dimension are isomorphic as Lie algebras. So now we know that the connected group G and the simply connected group $\mathbb{R}^n$ have isomorphic Lie algebras. By part (a) of Theorem 5.10*, this tells us that the additive group $\mathbb{R}^n$ is the universal cover of G. By part (b) of Theorem 5.10*, it follows that G is a homomorphic image of $\mathbb{R}^n$; that is, there is an onto homomorphism $F : \mathbb{R}^n \to G$. We want to show that G is abelian. Choose $M_1, M_2 \in G$. Because F is onto, there are vectors $\boldsymbol{v}_1$ and $\boldsymbol{v}_2$ in $\mathbb{R}^n$ with $F(\boldsymbol{v}_j) = M_j$. Now, because F is a homomorphism and $\mathbb{R}^n$ is an abelian group, we have

$$M_1 * M_2 = F(\boldsymbol{v}_1) * F(\boldsymbol{v}_2) = F(\boldsymbol{v}_1 + \boldsymbol{v}_2) = F(\boldsymbol{v}_2 + \boldsymbol{v}_1) = F(\boldsymbol{v}_2) * F(\boldsymbol{v}_1) = M_2 * M_1.$$

Therefore G is abelian. □

We include two other beautiful and important theorems.

Theorem 5.12* *Suppose G_1 and G_2 are Lie groups with G_1 simply connected. Suppose also that $f : L(G_1) \to L(G_2)$ is a Lie algebra homomorphism. Then there is a Lie group homomorphism $F : G_1 \to G_2$ with $f = dF$.* (Compare Theorem 3.9*.)

Theorem 5.13* (Main Theorem). *If G_1 and G_2 are simply connected Lie groups, and if their Lie algebras $L(G_1)$ and $L(G_2)$ are isomorphic as Lie algebras, then G_1 and G_2 are isomorphic as Lie groups.*

The "Lie theory" of Lie groups and Lie algebras encompasses much, much more than this. For example, classifying the "simple" Lie algebras yields a classification of the "nice" Lie groups. (The definition of a simple Lie algebra is in section 6.5.2.) There are four infinite families of simple Lie algebras, called the *classical algebras* and labeled A, B, C, D. We have seen all four of these families:

Family A consists of the n by n trace zero matrices. We encountered these as the Lie algebras of the special linear groups $SL(n, \mathbb{R})$.

Families B and D consist of n by n skew symmetric matrices (family B for odd n and family D for even n). These arise as the Lie algebras of the Euclidean orthogonal groups $O(n, \mathbb{R})$.

Family C consists of the $2m$ by $2m$ matrices of the form

$$\begin{bmatrix} M_{11} & M_{12} \\ M_{21} & M_{11}^T \end{bmatrix}$$

where the M_{ij} are m by m matrices, and M_{12} and M_{21} are symmetric (see (4.5.7)). These arise as the Lie algebras of the symplectic groups $Sp(2m, \mathbb{R})$.

The tangent spaces of the unitary groups (determined in Chapter 6) form an infinite family of Lie algebras too, but this family is a "twist" on family A. There is a twisted version of family D too. In addition, there are five exceptional simple Lie algebras, of dimensions 14, 52, 78, 133 and 248; three of these also have twisted versions. We'll describe a bit more about the classification of simple Lie algebras and the implications for the classification of "simple" groups in Chapter 6.

CHAPTER 6

Matrix groups over other fields

6.1 What is a field?

What happens when we consider matrices and vector spaces associated with number systems other than the real numbers? In order that the results from linear algebra on which our work depends remain valid, we require that the number system we use be a field. A **field** is a set $\mathbb{F}$ with two operations, addition and multiplication (for $a, b \in \mathbb{F}$ we write $a + b$ and ab for the results of these two operations) satisfying the following axioms.

Addition The elements of $\mathbb{F}$ form an abelian group under addition; the identity of this group is written 0 and called zero.

Multiplication The nonzero elements of $\mathbb{F}$ form an abelian group under multiplication; the identity of this group is written 1.

Distributive law The operations of addition and multiplication are linked by the distributive law: $a(b + c) = ab + ac$ for all $a, b, c \in \mathbb{F}$.

Examples of fields include the real and complex numbers, as well as the rational numbers with the usual addition and multiplication. (We'll look at more examples of fields in section 2.)

Problem 6.1.1.

a. The set $\mathbb{Z}$ of integers, with the usual addition and multiplication, isn't a field. What goes right and what goes wrong?

b. The set $\mathbb{H}$ of quaternions, with addition and multiplication defined as in section 2.2, isn't a field. What goes right and what goes wrong?

Using numbers from a field, we can carry out all the arithmetic operations we use for linear algebra (and other purposes). In particular, notice that in a field $\mathbb{F}$, the following familiar arithmetic properties are valid.

- It follows from the distributive law that $0 \cdot a = 0$ and $(-1) \cdot a = -a$ for any $a \in \mathbb{F}$. (We write $-a$ for the additive inverse of a in $\mathbb{F}$; we know it exists since $\mathbb{F}$ is a group under addition.)

Problem 6.1.2. (S)
Verify the preceding statements.

- We can regard every integer as an element of $\mathbb{F}$ as follows. We identify the integers 0 and 1 with the additive and multiplicative identities respectively in $\mathbb{F}$. Similarly, we identify the integer -1 with the additive inverse of 1 in $\mathbb{F}$. If m is a positive integer, then we identify m with the field element $1 + \cdots + 1$ (m terms). If $n = -m$ is a negative integer, then we identify n with the field element $(-1) + \cdots + (-1)$ (m terms).

Problem 6.1.3.
Assume $\mathbb{F}$ is a field, $a \in \mathbb{F}$, and m is a positive integer. Using the identifications above, explain why the following statements are true.
a. $ma = a + \cdots + a \quad$ (m terms).
b. $(-m)a = (-a) + \cdots + (-a) \quad$ (m terms).
c. $(-m)a = -(ma)$.

- We can subtract: $a - b$ means $a + (-b)$, which is an element of $\mathbb{F}$ when $a, b \in \mathbb{F}$.
- We can divide by any nonzero element of $\mathbb{F}$: If $a, b \in \mathbb{F}$ and $b \neq 0$, then (because the nonzero elements of $\mathbb{F}$ form a group under multiplication) the multiplicative inverse b^{-1} of b is also in $\mathbb{F}$, and a/b means ab^{-1}, which is an element of $\mathbb{F}$.

For any field $\mathbb{F}$ we can apply facts from linear algebra to matrices with entries from $\mathbb{F}$ and to the vector space $\mathbb{F}^n$. For example, matrix addition and multiplication have their usual properties, the determinant and trace are defined as usual and have their familiar properties, and a linear transformation $T : \mathbb{F}^n \to \mathbb{F}^n$ can be described by a matrix $M \in \mathcal{M}(n, \mathbb{F})$, with $T(\boldsymbol{x}) = M\boldsymbol{x}$ for all $\boldsymbol{x} \in \mathbb{F}^n$. Further, T is invertible if and only if $\det(M) \neq 0$. We obtain immediately that

$$GL(n, \mathbb{F}) = \{M \in \mathcal{M}(n, \mathbb{F}) : \det(M) \neq 0\} \quad \text{and}$$
$$SL(n, \mathbb{F}) = \{M \in GL(n, \mathbb{F}) : \det(M) = 1\}$$

are groups under matrix multiplication.

6.2 The unitary group

The complex numbers are different from some other fields in having an extra operation, complex conjugation. In Chapter 2 we defined two groups of matrices with entries in the complex numbers with criteria that depend on complex conjugation. We defined the **unitary group** $U(n, \mathbb{C})$,

$$U(n, \mathbb{C}) = \{M \in \mathcal{M}(n, \mathbb{C}) : \overline{M}^T M = I_n\},$$

and its subgroup the **special unitary group** $SU(n, \mathbb{C})$,

$$SU(n, \mathbb{C}) = \{M \in U(n, \mathbb{C}) : \det(M) = 1\}.$$

Our treatment of these groups in Chapter 2 was completely algebraic, but there is also some underlying geometry. The unitary group is the group of symmetries of the vector space $\mathbb{C}^n$ with a special dot product, called the *Hermitian* product. To motivate the definition of this new dot product, recall two facts.

- The Euclidean dot product on $\mathbb{R}^n$ describes distance and length—specifically, the length $|\boldsymbol{x}|$ of $\boldsymbol{x} \in \mathbb{R}^n$ is $\sqrt{\boldsymbol{x} \cdot \boldsymbol{x}}$.
- The length of a complex number $z = a + bi \in \mathbb{C}$ is $|z| = \sqrt{\overline{z}z} = \sqrt{a^2 + b^2}$.

For vectors $\boldsymbol{x} = (x_1, x_2, \dots, x_n)$ and $\boldsymbol{y} = (y_1, y_2, \dots, y_n)$ in $\mathbb{C}^n$, the **Hermitian product** of $\boldsymbol{x}$ and $\boldsymbol{y}$ is the complex number

$$\boldsymbol{x} \cdot \boldsymbol{y} = \overline{\boldsymbol{x}}^T \boldsymbol{y} = \overline{x_1} y_1 + \overline{x_2} y_2 + \cdots + \overline{x_n} y_n.$$

More formally, this dot product is called a **Hermitian form** on $\mathbb{C}^n$. In this section, the dot product on $\mathbb{C}^n$ is assumed to be the Hermitian product. (Have a look online at biographical information about the French mathematician Charles Hermite, 1822-1901. His name is attached to this dot product because of its association to what is called a *Hermitian* matrix: a matrix $A \in \mathcal{M}(n, \mathbb{C})$ is Hermitian if $A^T = \overline{A}$. This idea is related to (6.2.2c) below.)

Problem 6.2.1.
Verify that the complex numbers form a field. Cite results from chapter 2 as needed.

Problem 6.2.2.
Fix a positive integer n. Verify that the Hermitian dot product has the following properties for all vectors $\boldsymbol{x}$, $\boldsymbol{y}$, and $\boldsymbol{z}$ in $\mathbb{C}^n$ and all scalars $c \in \mathbb{C}$. Cite results from chapter 2 as needed.

a. Distributive laws:

$$\boldsymbol{x} \cdot (\boldsymbol{y} + \boldsymbol{z}) = \boldsymbol{x} \cdot \boldsymbol{y} + \boldsymbol{x} \cdot \boldsymbol{z}$$
$$(\boldsymbol{x} + \boldsymbol{y}) \cdot \boldsymbol{z} = \boldsymbol{x} \cdot \boldsymbol{z} + \boldsymbol{y} \cdot \boldsymbol{z}$$

b. Scalar multiplication:

$$\boldsymbol{x} \cdot (c\boldsymbol{y}) = c(\boldsymbol{x} \cdot \boldsymbol{y})$$
$$(c\boldsymbol{x}) \cdot \boldsymbol{y} = \overline{c}(\boldsymbol{x} \cdot \boldsymbol{y})$$

c. Semi-symmetry:

$$\boldsymbol{y} \cdot \boldsymbol{x} = \overline{\boldsymbol{x} \cdot \boldsymbol{y}}$$

d. **(S)** Non-degeneracy: If $\boldsymbol{v} \in \mathbb{C}^n$ has the property that $\boldsymbol{v} \cdot \boldsymbol{w} = 0$ for all $\boldsymbol{w} \in \mathbb{C}^n$, then $\boldsymbol{v}$ is the zero vector.

e. The Hermitian dot product $\boldsymbol{x} \cdot \boldsymbol{x}$ of a vector with itself is a real number.

f. The real number $\boldsymbol{x} \cdot \boldsymbol{x} \geq 0$, and $\boldsymbol{x} \cdot \boldsymbol{x} = 0$ if and only if $\boldsymbol{x}$ is the zero vector.

g. If all of the entries x_j and y_j of the vectors $\boldsymbol{x}, \boldsymbol{y} \in \mathbb{C}^n$ are real numbers, then the Hermitian product $\boldsymbol{x} \cdot \boldsymbol{y}$ agrees with the Euclidean dot product on $\mathbb{R}^n$.

From parts (e) and (f) of (6.2.2) we see that it is reasonable to define the **length** of a vector $v \in \mathbb{C}^n$ by $|v| = \sqrt{v \cdot v}$, and in the special case when $n = 1$, this agrees with our definition of the length of a complex number. In the next problem you will show that a matrix M is in the unitary group if and only if the linear transformation $T : \mathbb{C}^n \to \mathbb{C}^n$ defined by $T(v) = Mv$ for $v \in \mathbb{C}^n$ preserves the Hermitian form on $\mathbb{C}^n$. Following our notational convention in the preceding chapters, we write

$$U(\mathbb{C}^n) = \{T : \mathbb{C}^n \to \mathbb{C}^n \ : \ T \text{ linear and } T(x) \cdot T(y) = x \cdot y \text{ for all } x, y \in \mathbb{C}^n\}.$$

Problem 6.2.3.

Fix a positive integer n.

a. Show that a matrix $M \in \mathcal{M}(n, \mathbb{C})$ describes a linear transformation preserving the Hermitian dot product on $\mathbb{C}^n$ if and only if $\overline{M}^T M = I_n$.

b. Show $U(\mathbb{C}^n)$ is a group under composition of functions.

c. Explain why $U(\mathbb{C}^n)$ and $U(n, \mathbb{C})$ are isomorphic groups.

Remark Perhaps surprisingly, the mathematical setting for quantum mechanics is a complex vector space with a Hermitian dot product. The state of a particular physical system is represented by a vector v of length 1 (a unit vector) in a particular space. One of the special features of quantum mechanics is that a system can be in a "superposition" of several states, meaning a linear combination. For example, suppose the system is a particle with two spin states: the unit vector v_0 represents the state "spin up" and the unit vector v_1 represents the state "spin down." If the particle is in the state represented by the unit vector $z = av_0 + bv_1$ (so $z \cdot z = \bar{a}a + \bar{b}b = 1$), then measuring its state gives the result "spin up" with probability $\bar{a}a$ and "spin down" with probability $\bar{b}b$. The transition from one state to another is via a unitary transformation of the complex space. Recently, there has been much interest in the possibility of constructing a *quantum computer*, a computer that works by the laws of quantum mechanics rather than those of classical Newtonian physics. The *gates* of the quantum computer would be unitary transformations of the complex vector space containing the possible physical states of the computer. (For an elementary exposition of some of these ideas, see sections 1 and 2.1 of my paper in the *Monthly*, "Quantum Error Correction: classic group theory meets a quantum challenge.")

As we did for matrix groups over the real numbers, we want to associate a tangent space at the identity to each of these matrix groups over the complex numbers. In order to do so, we need some preliminaries.

Suppose $\gamma : \mathbb{R} \to \mathcal{M}(n, \mathbb{C})$. As usual, define

$$\gamma'(t) = \lim_{h \to 0} \frac{\gamma(t+h) - \gamma(t)}{h}.$$

Just as in the real case, if the entries of $\gamma(t)$ are $\gamma_{jk}(t)$ for $\gamma_{jk} : \mathbb{R} \to \mathbb{C}$, the entries of $\gamma'(t)$ are $\gamma'_{jk}(t)$. So, now we need to know how to differentiate a function $f : \mathbb{R} \to \mathbb{C}$. Again

$$f'(t) = \lim_{h \to 0} \frac{f(t+h) - f(t)}{h}.$$

If $f(t) = a(t) + ib(t)$, it follows from the definition that $f'(t) = a'(t) + ib'(t)$.

Problem 6.2.4.

a. As above, define $f : \mathbb{R} \to \mathbb{C}$ by $f(t) = a(t)+b(t)i$, where a and b are differentiable functions $\mathbb{R} \to \mathbb{R}$. Show $\overline{f'(t)} = (\overline{f(t)})'$; that is, show that you get the same result whether you differentiate first and then form the complex conjugate or form the complex conjugate first and then differentiate.

b. Fix a positive integer n. Assume $\gamma : \mathbb{R} \to \mathcal{M}(n, \mathbb{C})$, and the entries $\gamma_{ij}(t)$ are differentiable functions from $\mathbb{R}$ to $\mathbb{C}$, as in (a). Explain why $\overline{\gamma'(t)} = (\overline{\gamma(t)})'$.

As was true for functions with values in $\mathcal{M}(n, \mathbb{R})$, if α and β are differentiable functions $\mathbb{R} \to \mathcal{M}(n, \mathbb{C})$, then the product rule holds for differentiation with respect to t:

$$(\alpha(t)\beta(t))' = \alpha(t)\beta'(t) + \alpha'(t)\beta(t).$$

Problem 6.2.5.

With notation as above, verify the product rule for the special case where α and β are differentiable functions $\mathbb{R} \to \mathcal{M}(2, \mathbb{C})$.

Assume that the product rule holds for differentiable functions $\alpha, \beta : \mathbb{R} \to \mathcal{M}(n, \mathbb{C})$ for any positive integer n in what follows. Also, *assume* that $\gamma(0) = I_n$ and $\det(\gamma(t)) = 1$ imply that the trace of $\gamma'(0)$ is zero, as is true when we work over the real numbers. Now we can define a **smooth curve passing through the identity** in $GL(n, \mathbb{C})$ exactly as we did in the real case. For a subgroup G of $GL(n, \mathbb{C})$, define the **tangent space** of G by

$$L(G) = \{\gamma'(0) : \gamma : \mathbb{R} \to G, \gamma \text{ smooth, and } \gamma(0) = I_n\}.$$

Problem 6.2.6.

Fix a positive integer n. Suppose $G = SU(n, \mathbb{C})$ and $\gamma : \mathbb{R} \to G$ is a smooth curve passing through the identity.

a. **(S)** Show that $\gamma(t) \in U(n, \mathbb{C})$ implies

$$\overline{\gamma'(0)}^T + \gamma'(0) = 0_n.$$

b. Suppose $n = 2$ and

$$\gamma'(0) = \begin{bmatrix} a'(0) & b'(0) \\ c'(0) & d'(0) \end{bmatrix},$$

with $a'(0) = a_1 + i a_2$, etc. What does the condition in (a) tell you about the entries of $\gamma'(0)$?

c. Continue to assume $n = 2$. Show

$$L(G) \subseteq W = \left\{ \begin{bmatrix} ui & v + wi \\ -v + wi & -ui \end{bmatrix} : u, v, w \in \mathbb{R} \right\}.$$

d. Continuing the analysis of (c), show that W is closed under addition.

e. Again continuing (c), show that W is closed under multiplication by *real* scalars, but *not* closed in general under multiplication by complex scalars. (Give a specific counterexample.)

We see from (6.2.6e) that although we have broadened our analysis to vector spaces and groups over the complex numbers, the tangent space of a complex matrix group might be only a *real* vector space, not a complex one. Nonetheless, we carry on in extending our ideas from real matrix groups to complex ones.

The next step is to define the exponential function for complex matrices, which we can do exactly as we did in the real case. For $A \in \mathcal{M}(n, \mathbb{C})$, let

$$\exp(A) = I_n + A + \frac{1}{2!}A^2 + \frac{1}{3!}A^3 + \cdots + \frac{1}{m!}A^m + \cdots.$$

Assume that this series converges for all such A. In fact, the exponential function for complex matrices has all the properties in 4.4. Assume these facts as well.

Problem 6.2.7. (S)

Find $\exp(tA)$ for the following matrices.

a. $A = \begin{bmatrix} i & 0 \\ 0 & 0 \end{bmatrix}$, b. $A = \begin{bmatrix} 0 & i \\ 0 & 0 \end{bmatrix}$.

The next problem completes the determination of the tangent space of $G = SU(2, \mathbb{C})$.

Problem 6.2.8.

Continue the notation of (6.2.6). Let σ_1, σ_2, σ_3 be the *Pauli spin matrices*,

$$\sigma_1 = \begin{bmatrix} 0 & 1 \\ 1 & 0 \end{bmatrix}, \quad \sigma_2 = \begin{bmatrix} 0 & -i \\ i & 0 \end{bmatrix}, \quad \sigma_3 = \begin{bmatrix} 1 & 0 \\ 0 & -1 \end{bmatrix}.$$

Let $A_j = i\sigma_j$ for $j = 1, 2, 3$.

a. **(S)** Find $\exp(tA_1)$ and show it's in G.
b. Find $\exp(tA_2)$ and show it's in G.
c. Find $\exp(tA_3)$ and show it's in G.
d. Explain why (a)-(c) tell you that A_1, A_2 and A_3 are in $L(G)$.
e. Show that every element of W can be written as a real linear combination of A_1, A_2, A_3.
f. Explain why $L(G) = W$.

Problem 6.2.9.

Fix a positive integer n and let $A \in \mathcal{M}(n, \mathbb{C})$. Explain why $\exp(\overline{A}) = \overline{\exp(A)}$.

Problem 6.2.10.

Fix a positive integer n. Assume (3.3.12) and (4.5.6) hold for matrices with complex entries; that is, the differential of the determinant map is the trace, and $\det(\exp(A)) = \exp(\operatorname{tr}(A))$ for all $A \in \mathcal{M}(n, \mathbb{C})$. Let $G = U(n, \mathbb{C})$. By (6.2.6a), if $\gamma : \mathbb{R} \to G$ is a smooth curve passing through the identity and if $A = \gamma'(0)$, then $\overline{A}^T + A = 0_n$.

a. Assume $A \in \mathcal{M}(n, \mathbb{C})$ and $\overline{A}^T + A = 0_n$. Explain why $\overline{A}^T A = A\overline{A}^T$.
b. Let $G = U(n, \mathbb{C})$. Explain why $L(G) = \{A \in \mathcal{M}(n, \mathbb{C}) : \overline{A}^T + A = 0_n\}$.
c. Let $G = SU(n, \mathbb{C})$. Explain why

$$L(G) = \{A \in \mathcal{M}(n, \mathbb{C}) : \overline{A}^T + A = 0_n \text{ and } \operatorname{tr}(A) = 0\}.$$

d. Show that the tangent spaces in (c) and (d) are closed under the matrix bracket $[A, B] = AB - BA$.

e. Check that your answer in (c) agrees with your answer to (6.2.8) for the case $n = 2$.

Problem 6.2.11.

In Chapter 3 we saw that for G a subgroup of $GL(n, \mathbb{R})$, $L(G)$ is closed under addition. Adapt the argument from Chapter 3 to give the same conclusion for G a subgroup of $GL(n, \mathbb{C})$.

Problem 6.2.12.

In Chapter 3 we saw that for G a subgroup of $GL(n, \mathbb{R})$, $L(G)$ is closed under scalar multiplication. Adapt the argument from Chapter 3 for G a subgroup of $GL(n, \mathbb{C})$ to show that $A \in L(G)$ and $c \in \mathbb{R}$ implies $cA \in L(G)$. That is, show $L(G)$ is closed under multiplication by *real* scalars. What goes wrong in the argument if $c \in \mathbb{C}$ but c isn't a real number?

From (6.2.11) and (6.2.12) we know that for any subgroup G of $GL(n, \mathbb{C})$, we can say $L(G)$ is a vector space over $\mathbb{R}$, but not necessarily a vector space over $\mathbb{C}$. Indeed, (6.2.6) and (6.2.8) show that for $G = SU(2, \mathbb{C})$, $L(G)$ is *not* closed under multiplication by complex scalars.

Problem 6.2.13.

Fix a positive integer n. Explain why the following statements are true.

a. If $G = GL(n, \mathbb{C})$, then $L(G) = \mathcal{M}(n, \mathbb{C})$.

b. If $G = SL(n, \mathbb{C})$ then

$$L(G) = \{A \in \mathcal{M}(n, \mathbb{C}) \; : \; \mathrm{tr}(A) = 0\}.$$

c. The tangent spaces in (a) and (b) are closed under multiplication by complex scalars.

d. The tangent spaces in (a) and (b) are closed under the matrix bracket $[A, B] = AB - BA$.

6.3 Matrix groups over finite fields

The real numbers $\mathbb{R}$ and the complex numbers $\mathbb{C}$ are *infinite* fields. But fields can be finite sets. The world's smallest field is $\mathbb{F}_2 = \{0, 1\}$ with addition and multiplication defined by

$$\begin{array}{c|cc} + & 0 & 1 \\ \hline 0 & 0 & 1 \\ 1 & 1 & 0 \end{array} \quad \text{and} \quad \begin{array}{c|cc} \cdot & 0 & 1 \\ \hline 0 & 0 & 0 \\ 1 & 0 & 1 \end{array} .$$

Problem 6.3.1.

a. Verify that $\mathbb{F}_2$ is a field. (You can assume that the associative and distributive laws hold.)

b. Show that $x + x = 0$ and $x^2 = x$ for every x in $\mathbb{F}_2$.

In fact, there are lots of finite fields. More examples can be created by generalizing the example of $\mathbb{F}_2$ to $\mathbb{F}_p$ for any prime p. The elements of $\mathbb{F}_p$ are the integers which are the distinct nonnegative proper remainders on division of an integer by p,

$$\mathbb{F}_p = \{0, 1, 2, \ldots, p-1\}.$$

Define addition and multiplication **modulo** p: for $a, b \in \mathbb{F}_p$, if $a + b$ or ab computed using ordinary arithmetic exceeds $p - 1$, replace the resulting integer by its nonnegative proper remainder on division by p. For example, if $p = 5$, then $1+2 = 3$ but $4+2 = 1$ and $4 + 1 = 0$. Similarly, working modulo 5, $2 \cdot 2 = 4$ and $2 \cdot 3 = 1$. It can be shown that these definitions make $\mathbb{F}_p$ a field for any choice of the prime p.

Problem 6.3.2. (S)

a. Write out the addition and multiplication tables for $\mathbb{F}_5$.

b. Use your tables to verify that $\mathbb{F}_5$ is a field. (You can assume that the associative and distributive laws hold.)

c. What is the value of $3 - 4$ in this field? of $3/4$?

d. We saw in section 1 that given a particular field, every integer can be regarded as an element of that field. How do we interpret the integer 5 as an element of the field $\mathbb{F}_5$? What is the interpretation of the integer 17 as an element of $\mathbb{F}_5$? the interpretation of -17?

Problem 6.3.3.

What happens if you try the construction described above using modular arithmetic but working with a composite number rather than a prime? Specifically, suppose you try to construct a field using arithmetic modulo 6, working with the numbers $\mathbb{S} = \{0, 1, 2, 3, 4, 5\}$ and defining addition and multiplication modulo 6.

a. Write out the addition and multiplication tables for $\mathbb{S}$.

b. Is $\mathbb{S}$ a field? What goes right and what goes wrong? (You can assume that the associative and distributive laws hold.) (Note: If you've studied abstract algebra, you may recognize that $\mathbb{S}$ is a *ring*, often denoted $\mathbb{Z}/6\mathbb{Z}$.)

Problem 6.3.4.

Assume $\mathbb{F}$ is a field. The Heisenberg group $\mathcal{H}$ is defined by

$$\mathcal{H} = \left\{ \begin{bmatrix} 1 & a & b \\ 0 & 1 & c \\ 0 & 0 & 1 \end{bmatrix} : a, b, c \in \mathbb{F} \right\}.$$

We have seen this group before for the special case $\mathbb{F} = \mathbb{R}$ in (5.1.10); the same argument shows that $\mathcal{H}$ is a group under matrix multiplication for any field.

a. Assume $\mathbb{F}$ is a field in which the integer 2 is interpreted as a nonzero element of the field (so division by 2 makes sense in $\mathbb{F}$). Use induction to show that for any $a, b, c \in \mathbb{F}$,

$$\begin{bmatrix} 1 & a & b \\ 0 & 1 & c \\ 0 & 0 & 1 \end{bmatrix}^n = \begin{bmatrix} 1 & na & nb + n(n-1)ac/2 \\ 0 & 1 & nc \\ 0 & 0 & 1 \end{bmatrix}$$

for all positive integers n.

For the remaining parts of this problem, assume that the field $\mathbb{F} = \mathbb{F}_p$, the field constructed using arithmetic modulo p for some prime p.

b. Assume $p > 2$ and use (a) to show that $A \in \mathcal{H}$ implies $A^p = I_3$.

c. Explain why there are p^3 matrices in $\mathcal{H}$.

d. If $p = 2$, is it true that $A \in \mathcal{H}$ implies $A^2 = I_3$? Why or why not?

Remark The Heisenberg group has many applications, and they vary depending on what the field $\mathbb{F}$ is. The version of the Heisenberg group when the field $\mathbb{F}$ is the real numbers is used in quantum mechanics. The version when $\mathbb{F} = \mathbb{C}$ and the off-diagonal elements a, b, c satisfy additional conditions is used in an area of mathematics called harmonic analysis. The version when the field $\mathbb{F}$ is $\mathbb{F}_p$ for an odd prime p is an example of what group theorists call an *extra special p-group*.

Extra special 2-groups also exist, although this construction doesn't produce one. They are used in the theory of error correction for a quantum computer. The smallest extra-special 2-group has 8 elements. One description is as the following group of 2 by 2 matrices (often called the *Pauli group*), $\mathcal{P} = \{I_2, -I_2, X, -X, Z, -Z, Y = XZ, -Y = ZX\}$, where

$$X = \begin{bmatrix} 0 & 1 \\ 1 & 0 \end{bmatrix} \quad \text{and} \quad Z = \begin{bmatrix} 1 & 0 \\ 0 & -1 \end{bmatrix}$$

are the Pauli spin matrices σ_1 and σ_3 of (6.2.8). The symplectic and orthogonal groups $Sp(n, \mathbb{F}_2)$ and $O(n, \mathbb{F}_2)$ also play roles in the theory of extra-special 2-groups and thus in the mathematics of error correction for quantum computers.

In the rest of this section, we will explore the finite group $G = SL(3, \mathbb{F}_2)$ as a group of symmetries of the vector space $V = (\mathbb{F}_2)^3$ and a related geometry called a *symmetric design*. (There are 168 matrices in G, so you certainly won't list them all!)

The vectors in V have the form $\boldsymbol{v} = (v_1, v_2, v_3)$ with $v_j \in \mathbb{F}_2$, and addition and scalar multiplication of vectors are defined using the addition and multiplication rules for elements of $\mathbb{F}_2$. Because there are two choices for each coordinate of a vector in V, V contains a total of $2^3 = 8$ vectors. Since the only scalars are 0 and 1, every 1-dimensional subspace of V consists of the zero vector and one nonzero vector. Consequently, two distinct nonzero vectors are necessarily linearly independent. There are thus exactly seven 1-dimensional subspaces in V.

A 2-dimensional subspace of V has two basis vectors $\boldsymbol{v}$ and $\boldsymbol{w}$, and it consists of all vectors of the form $a\boldsymbol{v} + b\boldsymbol{w}$ where a and b vary over $\mathbb{F}_2$. Since there are two choices for a and two choices for b, each 2-dimensional subspace contains $2^2 = 4$ vectors, three of them nonzero. For example, $\boldsymbol{v} = (1, 0, 0)$ and $\boldsymbol{w} = (1, 1, 0)$ are linearly independent, and the nonzero vectors in the subspace spanned by $\boldsymbol{v}$ and $\boldsymbol{w}$ are

$$\boldsymbol{v} = (1, 0, 0) \quad \boldsymbol{w} = (1, 1, 0) \quad \boldsymbol{v} + \boldsymbol{w} = (0, 1, 0).$$

Problem 6.3.5. (S)

a. List the seven nonzero vectors in V. (Don't bother to give names to the vectors.)

b. For each of the seven 2-dimensional subspaces of V, list the three nonzero vectors it contains, as in the preceding example.

It's possible to draw a nice picture of the seven 1-dimensional subspaces (i.e., the seven nonzero vectors) of V in a way that shows the triples of nonzero vectors in each 2-dimensional subspace. Each 1-dimensional subspace (nonzero vector) is represented as a *point*, and it is labeled by the coordinates of the vector. Each triple of points belonging to the same 2-dimensional subspace is called a *block*; there are seven blocks. Six of the seven 2-dimensional subspaces are represented by line segments; one 2-dimensional subspace is represented by a circle. A point is drawn on a line segment or circle if the corresponding vector belongs to the corresponding 2-dimensional subspace, so there are three points on each block. The picture of the seven points and seven blocks is shown in Figure 6.1 below. It has been partially labeled. (Parentheses and commas have been omitted in the labels.)

Problem 6.3.6.
Complete the labeling of Figure 6.1.

Figure 6.1 illustrates an example of a (v, k, λ) *symmetric design*, with $v = 7$, $k = 3$ and $\lambda = 1$. The design consists of $v = 7$ *points* and $v = 7$ *blocks*. Each block consists of the $k = 3$ points belonging to the same 2-dimensional subspace, and each point belongs to $k = 3$ blocks. Two distinct blocks have exactly $\lambda = 1$ point in common, and two distinct points belong to exactly $\lambda = 1$ block. This particular design is also an example of a *projective plane*: two points determine a unique line (i.e., block), and every pair of lines meets at a unique point. Notice there are, therefore, no parallel lines in this geometry; it is an example of a *non-Euclidean* geometry. It is often called the *Fano plane*.

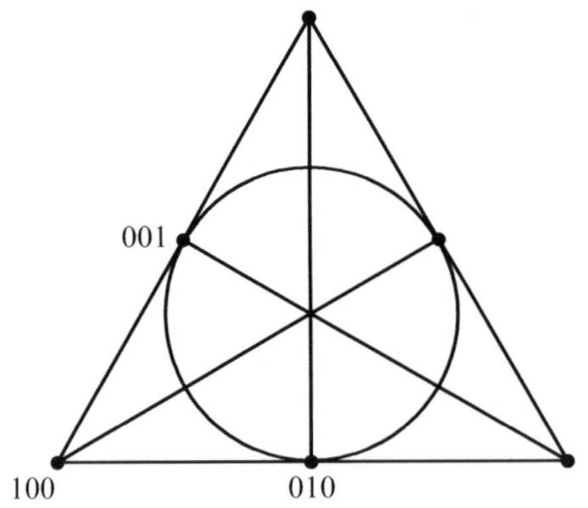

Figure 6.1. A picture of a $(7, 3, 1)$ symmetric design called the Fano plane

Problem 6.3.7. (S)
Let

$$M_1 = \begin{bmatrix} 1 & 0 & 0 \\ 0 & 0 & 1 \\ 0 & 1 & 0 \end{bmatrix}.$$

a. Verify that M_1 is in $G = SL(3, \mathbb{F}_2)$. (Remember to use the arithmetic of $\mathbb{F}_2$ in calculating the determinant.)

b. Calculate $(M_1)^2$.

c. The linear transformation $T_1 : V \to V$ corresponding to M_1 takes nonzero vectors to nonzero vectors (since M_1 is invertible). There is a simple description for the effect of T_1 on the picture of the Fano plane in Figure 6.1. It corresponds to a symmetry of the triangle. What symmetry?

Problem 6.3.8.
Let

$$M_2 = \begin{bmatrix} 1 & 0 & 0 \\ 1 & 1 & 1 \\ 0 & 1 & 0 \end{bmatrix}.$$

a. Verify that M_2 is in $G = SL(3, \mathbb{F}_2)$.

b. Calculate $(M_2)^3$.

c. The linear transformation $T_2 : V \to V$ corresponding to M_2 takes nonzero vectors to nonzero vectors (since M_2 is invertible). There is a simple description for the effect of T_2 on the picture in Figure 6.1. It corresponds to a symmetry of the triangle. What symmetry?

Problem 6.3.9.
Let

$$M_3 = \begin{bmatrix} 0 & 1 & 1 \\ 1 & 0 & 0 \\ 0 & 1 & 0 \end{bmatrix}.$$

a. Verify that M_3 is in $G = SL(3, \mathbb{F}_2)$.

b. Calculate $(M_3)^j$ for $j = 2, 3, \ldots, 7$.

c. The linear transformation $T_3 : V \to V$ corresponding to M_3 takes nonzero vectors to nonzero vectors (since M_3 is invertible). However, it does *not* correspond to a symmetry of the triangle in Figure 6.1. But, since it maps blocks to blocks in the Fano plane, it does correspond to something else that is nice. The remaining parts of this problem set up that correspondence. Let $P_0 = (1, 0, 0)$, and for $j = 1, \ldots, 6$, let $P_j = (T_3)^j(P_0)$, where $(T_3)^j$ corresponds to the matrix $(M_3)^j$. Figure out what each P_j is. What would P_7 be?

d. Re-list the triples in your answer to (6.2.5b) using these labels. After suitable reordering, here is what you should get.

$$\begin{array}{lll} \{P_1 & P_2 & P_4\} \\ \{P_2 & P_3 & P_5\} \\ \{P_3 & P_4 & P_6\} \\ \{P_4 & P_5 & P_0\} \\ \{P_5 & P_6 & P_1\} \\ \{P_6 & P_0 & P_2\} \\ \{P_0 & P_1 & P_3\} \end{array}$$

What do you notice about this description? In this notation, what's the effect of T_3 on the seven nonzero vectors of V? What's the effect of T_3 on the seven blocks listed above?

e. There is a way to redraw the (7, 3, 1) symmetric design to exhibit the effect of T_3 as a symmetry, and it appears in Figure 6.2 below. In Figure 6.2, the 3 nonzero vectors in the same 2-dimensional subspace are connected by an arc, rather than by a line segment or a circle. What is the symmetry of the drawing in Figure 6.2 that corresponds to the effect of T_3?

Here is the new picture of the $(7, 3, 1)$ symmetric design. In this picture, point P_j is just labeled j.

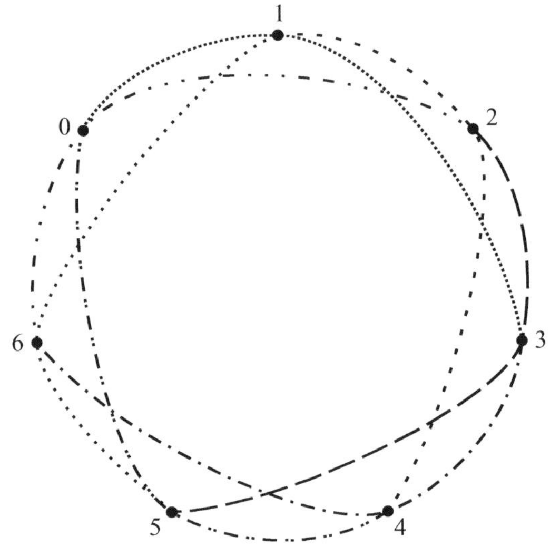

Figure 6.2. A drawing of the $(7, 3, 1)$ symmetric design that exhibits a symmetry T with $T^7 = I_3$

Remark If you have studied difference sets in cyclic groups, the subscripts in the list of points in (6.2.9d) might look familiar. We know that $\mathbb{F}_7 = \{0, 1, 2, 3, 4, 5, 6\}$ is a group under addition modulo 7. The elements in $D = \{1, 2, 4\}$ are the nonzero squares in $\mathbb{F}_7$. (Do you see why?) The elements in D form a (v, k, λ) *difference set* in the additive group $\mathbb{F}_7$: the list of all possible differences of distinct elements of D includes each nonzero group element exactly λ times. In this case $v = 7$ is the size of the group, $k = 3$ is the size of the set D, and $\lambda = 1$. Adding the same group element to each element of the difference set (using addition modulo 7) creates a new difference set. The seven difference sets obtainable from D in this way (one is D itself) correspond to the seven blocks listed in (6.3.9d).

It turns out that *every* (v, k, λ) difference set in a group is associated with a (v, k, λ) symmetric design. The v group elements are the v points of the design, the difference set and its translates by group elements are the v blocks. Each block contains k points (and each point belongs to k blocks). Two distinct blocks meet in λ points, and a specific pair of distinct points can be found in λ blocks. In fact, a group contains a difference set if and only if the group can be regarded as a "nice" group of symmetries of a symmetric design.

Symmetric designs have a number of applications. They are associated with classical error-correcting codes for protecting information from noise. They are also associated with experimental design in statistics (which is where the "v" comes from: it was originally for "varieties" of some crop plant in an agricultural study).

The loveliest book on symmetric designs that I know is *Symmetric Designs: an Algebraic Approach* by Eric Lander. Lander was trained as a pure mathematician—he earned his PhD in mathematics at Oxford. But if you search for Eric Lander online, you will find that he is now a geneticist. After teaching himself biology (!), Lander joined MIT as a tenured faculty member and a year later he launched the Whitehead Institute/MIT Center for Genome Research, becoming director of one of the first genome sequencing centers in the world. He recently founded and directs the joint Harvard-MIT Broad Institute for research in genomic medicine. You never know where mathematics might lead you.

You may feel that this section has stopped short: we have matrix groups over finite fields, but where are the corresponding Lie algebras? For an answer, see section 5.

6.4 Putting the pieces together

Definition A set $\mathbb{F}$ with two (binary) operations called addition and multiplication (denoted $a + b$ and ab respectively for $a, b \in \mathbb{F}$) is a **field** if:

Addition The elements of $\mathbb{F}$ form an abelian group under addition; the identity of this group is zero, denoted by 0 as usual.

Multiplication The nonzero elements of $\mathbb{F}$ form an abelian group under multiplication; the identity of this group is denoted by 1 as usual.

Distributive law The operations of addition and multiplication are linked by the distributive law: $a(b + c) = ab + ac$ for all $a, b, c \in \mathbb{F}$.

- The additive inverse of an element a in a field $\mathbb{F}$ is denoted by $-a$.
- In a field $\mathbb{F}$, subtraction is defined by $a - b = a + (-b)$ for $a, b \in \mathbb{F}$.
- The multiplicative inverse of a nonzero element b in a field $\mathbb{F}$ is denoted by b^{-1} or $1/b$.
- In a field $\mathbb{F}$, division is defined by $a/b = ab^{-1}$ for $a, b \in \mathbb{F}$ and b nonzero.
- For any field $\mathbb{F}$, every integer can be interpreted as an element of $\mathbb{F}$ as follows. The integers 0 and 1 are identified with the additive and multiplicative identities of $\mathbb{F}$. The positive integer m is identified with $1 + \cdots + 1$ (m terms), and the negative integer $-m$ is identified with $(-1) + \cdots + (-1)$ (m terms).

Examples

The rational numbers $\mathbb{Q}$, the real numbers $\mathbb{R}$, and the complex numbers $\mathbb{C}$ are all examples of fields, where addition and multiplication are defined as usual. Each of these fields contains an infinite number of elements.

There are also examples of *finite* sets which are fields under suitable definitions of addition and multiplication. A family of such examples can be obtained from the **integers modulo** p for a prime p. The elements are the distinct non-negative proper remainders when dividing an integer by p,

$$\mathbb{F}_p = \{0, 1, \ldots, p - 1\}.$$

Arithmetic is done *modulo* p: for $a, b \in \mathbb{F}_p$, if the value of ab or $a + b$ computed using ordinary arithmetic exceeds $p - 1$, replace the value by its proper nonnegative remainder on division by p. (If you've studied abstract algebra, you'll recognize that $\mathbb{F}_p \simeq \mathbb{Z}/p\mathbb{Z}$.)

The smallest possible field is $\mathbb{F}_2 = \{0, 1\}$, with addition and multiplication modulo 2. It's a theorem that every finite field $\mathbb{F}_q$ has $q = p^m$ elements for some prime p. Conversely, a field with p^m elements exists for every prime power p^m; it is unique up to a field isomorphism. Sometimes a finite field is called a *Galois field* and denoted by $GF(q)$ instead of $\mathbb{F}_q$.

Proposition 6.1 *If $\mathbb{F}$ is a field, then $0 \cdot a = 0$ for every $a \in \mathbb{F}$. Also, if $a \in \mathbb{F}$ and m is a positive integer, then, interpreting m and $-m$ as elements of $\mathbb{F}$, $(-m)a = -(ma)$.*

Linear algebra for vector spaces over the real numbers and matrices with real number entries can be adapted to any field. In particular

$$\mathbb{F}^n = \{(x_1, x_2, \ldots, x_n) : x_k \in \mathbb{F}\}$$

is a vector space with scalars in $\mathbb{F}$. Bases, linear transformations, matrices, matrix addition and multiplication, the determinant, and the trace are all defined as usual and have the usual properties.

When $\mathbb{F} = \mathbb{C}$, we have the complex analog of each of our matrix groups over the real numbers. We also have two matrix groups defined only over the complex numbers, namely the **unitary group**

$$U(n, \mathbb{C}) = \{M \in \mathcal{M}(n, \mathbb{C}) : \overline{M}^T M = I_n\},$$

and its subgroup $SU(n, \mathbb{C})$ of matrices of determinant 1. Although we defined these groups in Chapter 2, only in this chapter did we describe them as groups of symmetries of a vector space with a Hermitian dot product (or, more formally, a Hermitian form).

Definition Let $\boldsymbol{x} = (x_1, \ldots, x_n)$ and $\boldsymbol{y} = (y_1, \ldots, y_n)$ be two vectors in $\mathbb{C}^n$. The **Hermitian dot product** of $\boldsymbol{x}$ and $\boldsymbol{y}$ is the complex number

$$\boldsymbol{x} \cdot \boldsymbol{y} = \overline{x}_1 y_1 + \cdots + \overline{x}_n y_n.$$

Proposition 6.2 *The Hermitian dot product has the following properties for all vectors $\boldsymbol{x}$, $\boldsymbol{y}$, and $\boldsymbol{z}$ in $\mathbb{C}^n$ and all scalars $c \in \mathbb{C}$.*

- *Distributive laws*:

$$\boldsymbol{x} \cdot (\boldsymbol{y} + \boldsymbol{z}) = \boldsymbol{x} \cdot \boldsymbol{y} + \boldsymbol{x} \cdot \boldsymbol{z}$$
$$(\boldsymbol{x} + \boldsymbol{y}) \cdot \boldsymbol{z} = \boldsymbol{x} \cdot \boldsymbol{z} + \boldsymbol{y} \cdot \boldsymbol{z}$$

- *Scalar multiplication*:

$$\boldsymbol{x} \cdot (c\boldsymbol{y}) = c(\boldsymbol{x} \cdot \boldsymbol{y})$$
$$(c\boldsymbol{x}) \cdot \boldsymbol{y} = \overline{c}(\boldsymbol{x} \cdot \boldsymbol{y})$$

- *Semi-symmetry*:

$$\boldsymbol{y} \cdot \boldsymbol{x} = \overline{\boldsymbol{x} \cdot \boldsymbol{y}}$$

- *The Hermitian dot product* $\boldsymbol{x} \cdot \boldsymbol{x}$ *of a vector with itself is a real number;* $\boldsymbol{x} \cdot \boldsymbol{x} \geq 0$, *and* $\boldsymbol{x} \cdot \boldsymbol{x} = 0$ *if and only if* $\boldsymbol{x}$ *is the zero vector.*

- *If all of the entries* x_j *and* y_j *of the vectors* $\boldsymbol{x}, \boldsymbol{y} \in \mathbb{C}^n$ *are real numbers, then the Hermitian dot product* $\boldsymbol{x} \cdot \boldsymbol{y}$ *agrees with the Euclidean dot product on* $\mathbb{R}^n$.

We can think of the standard dot product on $\mathbb{R}^n$ as the restriction of the Hermitian dot product to real vectors. Similarly, we can think of the (Euclidean) orthogonal group $O(n, \mathbb{R})$ as a subgroup of $U(n, \mathbb{C})$.

Although the Hermitian dot product is not exactly bilinear (why not?), it is still true that the dot product is completely determined by its effect on basis vectors.

Proposition 6.3 *If* $\boldsymbol{e}_1, \ldots, \boldsymbol{e}_n$ *is a basis of* $\mathbb{C}^n$, *and if*

$$\boldsymbol{x} = a_1\boldsymbol{e}_1 + \cdots + a_n\boldsymbol{e}_n$$
$$\boldsymbol{y} = b_1\boldsymbol{e}_1 + \cdots + b_n\boldsymbol{e}_n$$

then we have

$$\boldsymbol{x} \cdot \boldsymbol{y} = \sum_{r,s=1}^{n} \overline{a}_r b_s \, (\boldsymbol{e}_r \cdot \boldsymbol{e}_s) \, .$$

Thus the Hermitian product can be specified by a basis $\boldsymbol{e}_1, \ldots, \boldsymbol{e}_n$ and an n by n matrix C whose r, s entry is the complex number $\boldsymbol{e}_r \cdot \boldsymbol{e}_s$. Notice that if the basis is the standard basis $(1, 0, \ldots, 0), \ldots, (0, 0, \ldots, 1)$, then the matrix of the Hermitian product is the identity matrix I_n.

Proposition 6.4 *The group* $U(n, \mathbb{C})$ *is isomorphic to*

$$U(\mathbb{C}^n) = \{T \in GL(n, \mathbb{C}) : T(\boldsymbol{x}) \cdot T(\boldsymbol{y}) = \boldsymbol{x} \cdot \boldsymbol{y} \text{ for all } \boldsymbol{x}, \boldsymbol{y} \in \mathbb{C}^n\}.$$

This group is also called the unitary group.

In order to define tangent vectors and tangent spaces, we have to be able to calculate $f'(t)$ when f is a complex-valued function of a real number:

$$f : \mathbb{R} \to \mathbb{C}.$$

As usual,

$$f'(t) = \lim_{h \to 0} \frac{f(t+h) - f(t)}{h}.$$

If we write $f(t) = a(t) + b(t)i$, where a and b are real-valued functions, then $f'(t) = a'(t) + b'(t)i$.

With this rule for differentiation, we find that our results for the real case carry over unchanged. (For example, the product rule continues to hold for derivatives of matrix-valued functions and matrix multiplication.) And differentiation relates simply to complex conjugation, as the following proposition shows.

Proposition 6.5 *Suppose the entries $a_{rs} = a_{rs}(t)$ of the n by m matrix A are differentiable complex-valued functions of the real variable t. Then*

$$\overline{\frac{\mathrm{d}A}{\mathrm{d}t}} = \frac{\mathrm{d}\overline{A}}{\mathrm{d}t}.$$

The definition of the exponential function can likewise be extended to matrices with complex entries,

$$\exp(A) = I + A + \frac{1}{2!}A^2 + \cdots + \frac{1}{m!}A^m + \cdots.$$

The exponential function for complex matrices has all the properties listed in 4.4. Therefore, we can find tangent spaces exactly as before. We also have the following.

Proposition 6.6 *If G is a subgroup of $GL(n, \mathbb{C})$, then its tangent space at the identity $L(G)$ is closed under addition and closed under multiplication by real scalars. (It is also closed under the matrix bracket, but we can't prove this in the general case using our methods.)*

Proposition 6.7 *The following is a partial list of complex matrix groups and their corresponding Lie algebras. (The description of the orthogonal groups assumes the Euclidean case.)*

Lie group	Lie algebra
$GL(n, \mathbb{C})$	$\mathcal{M}(n, \mathbb{C})$
$SL(n, \mathbb{C})$	$\{A \in \mathcal{M}(n, \mathbb{C}) : \mathrm{tr}(A) = 0\}$
$O(n, \mathbb{C}) = \{M \in GL(n, \mathbb{C}) : M^T M = I_n\}$	$\{A \in \mathcal{M}(n, \mathbb{C}) : A^T + A = 0_n\}$
$SO(n, \mathbb{C}) = \{M \in O(n, \mathbb{C}) : \det(M) = 1\}$	$\{A \in \mathcal{M}(n, \mathbb{C}) : A^T + A = 0_n\}$
$U(n, \mathbb{C}) = \{M \in \mathcal{M}(n, \mathbb{C}) : \overline{M}^T M = I_n\}$	$\{A \in \mathcal{M}(n, \mathbb{C}) : \overline{A}^T + A = 0_n\}$
$SU(n, \mathbb{C}) = \{M \in U(n, \mathbb{C}) : \det(M) = 1\}$	$\{A \in \mathcal{M}(n, \mathbb{C}) : \overline{A}^T + A = 0_n, \mathrm{tr}(A) = 0\}$

Proposition 6.8 *Let $\sigma_1, \sigma_2, \sigma_3$ be the Pauli spin matrices,*

$$\sigma_1 = \begin{bmatrix} 0 & 1 \\ 1 & 0 \end{bmatrix}, \quad \sigma_2 = \begin{bmatrix} 0 & -i \\ i & 0 \end{bmatrix}, \quad \sigma_3 = \begin{bmatrix} 1 & 0 \\ 0 & -1 \end{bmatrix}.$$

Then $A_1 = i\sigma_1$, $A_2 = i\sigma_2$, and $A_3 = i\sigma_3$ form a basis of the Lie algebra of $SU(2, \mathbb{C})$ as a real vector space.

Proposition 6.9 *The matrix groups $GL(n, \mathbb{F})$, $SL(n, \mathbb{F})$, $O(n, \mathbb{F})$, $SO(n, \mathbb{F})$, and $Sp\,(n, \mathbb{F})$ can all be defined over any field $\mathbb{F}$, including a finite field. (In the orthogonal or symplectic group case, as long as the entries of the matrix C defining the bilinear form are interpretable in $\mathbb{F}$ and C remains invertible when so interpreted, the criterion $M^T CM = C$ makes sense and defines a group. However, for a field, like $\mathbb{F}_2$, in which $-1 = +1$, the standard definitions of the orthogonal and symplectic groups are slightly different.) The same is true for the corresponding groups of linear transformations.*

Vector spaces over finite fields can be used to define symmetric designs. A **symmetric design** with parameters (v, k, λ) is a set of v *points*, and the points are grouped into v subsets of size k called *blocks* so that every point belongs to k blocks, two distinct points are together in λ blocks, and two distinct blocks have λ points in common. Although we explored only the case when $n = 3$, if n is any integer greater than or equal to 3 and if V is a vector space of dimension n over a finite field $\mathbb{F}$, then we can define a symmetric design by taking for the points of the design the 1-dimensional subspaces of V, and for the blocks the $(n-1)$-dimensional subspaces of V. It's always true that a linear transformation in $GL(n, \mathbb{F})$ maps points of this design to points, and maps blocks to blocks, and so produces a symmetry of the design. In the special case when $\mathbb{F} = \mathbb{F}_2$, the only possible nonzero determinant is 1, and $GL(n, \mathbb{F}_2) = SL(n, \mathbb{F}_2)$.

Proposition 6.10 *The group $SL(3, \mathbb{F}_2)$ is a group of symmetries of a $(7, 3, 1)$ symmetric design known as the Fano plane.*

Suggestions for further reading

Brown, E., "The many names of $(7, 3, 1)$," *Mathematics Magazine* **75** (2002) 83–94. This delightful article describes the Fano plane, difference sets and more.

Brown, E.,"The Fabulous $(11, 5, 2)$ Biplane," *Mathematics Magagazine* **77** (2004), 87–100. This article develops farther many of the themes in the preceding one.

Brown, E. and Mellinger, K.E., "Kirkman's Schoolgirls Wearing Hats and Walking through Fields of Numbers," *Mathematics Magazine* **82** (2009) 3–15. This is another excursion into designs.

Carter, R.W., *Simple Groups of Lie Type*, John Wiley and Sons (1972). The first three chapters (on the classical simple groups, Weyl groups and simple Lie algebras) are accessible to a determined and ambitious undergraduate reader.

The following three references are provided for ideas presented in the next section.

Humphreys, J.E., *Introduction to Lie Algebras and Representation Theory*, Graduate Texts in Mathematics **9**, Springer-Verlag (1972). Chapter VII contains a proof of the existence of a Chevalley basis and a description of the construction of Chevalley groups of adjoint type.

Ronan, M., *Symmetry and the Monster*, Oxford University Press (2006). This book tells the story of the classification of the finite simple groups at a level accessible to a general reader.

Thompson, T.M., *From Error-Correcting Codes through Sphere Packings to Simple Groups*, Mathematical Association of America (1983). This engaging book puts a very human face on mathematics. Its chapter 3 contains the story of J.H. Conway's discovery of three sporadic simple groups related to the symmetry group of a geometric object in 24 dimensions called the *Leech Lattice*. (The Leech Lattice is relevant to the sphere packings of the title.)

6.5 A broader view: finite groups of Lie type and simple groups

6.5.1 Finite groups of Lie type

We can define a variety of finite matrix groups by choosing the matrix entries from a finite field. These groups are of interest in their own right, but what about tangent spaces and Lie algebras? When we generalized from the real numbers to the complex numbers in section 2, we could still do calculus by expressing complex-valued functions in terms of familiar real-valued functions. The finite fields we have seen are intrinsically discrete, and it doesn't make sense to talk about limits as we must to do calculus. The Lie theory in this situation takes a completely different path. Instead of beginning with Lie groups and obtaining Lie algebras as tangent spaces, we begin with Lie algebras.

A Lie algebra L is a vector space (with scalars from a field $\mathbb{F}$) with a bracket operation satisfying certain axioms (as in 5.1). The study of Lie algebras makes sense working over *any* field, infinite or finite, although to get strong theorems one does need to make assumptions about $\mathbb{F}$, usually that $\mathbb{F}$ has *characteristic 0* (i.e., nonzero integers have nonzero value when interpreted as elements of $\mathbb{F}$) and is *algebraically closed*, assumptions that are satisfied by the complex numbers. Starting with a Lie algebra L over the complex numbers, it is possible to construct a matrix group $G = G(L, \mathbb{F})$ for any field $\mathbb{F}$ we choose. This matrix group is actually a group of "symmetries" of L, where by "symmetry" we mean a Lie algebra *automorphism* (i.e., a Lie algebra isomorphism from L to itself). In this roundabout way, we will come back to our finite matrix groups. The subject of finite groups of Lie type is immense. Although this *broader view* section is long compared to the preceding ones, we give only the briefest sketch of this immensity. This section is expository and contains no problems, but you should check your understanding of the reading by working through the two extended examples.

Proposition 6.5.1 *Let L be a Lie algebra (over any field), and let*

$$\mathrm{Aut}\,(L) = \{T : L \to L : T \textit{ is a Lie algebra automorphism}\}.$$

Then Aut (L) *is a group under composition of functions.*

Proof We know that the invertible linear transformations of L form a group. We just need to check that the composition of two linear transformations that preserve the bracket also preserves the bracket, and that the inverse of a linear transformation that preserves the bracket also preserves the bracket. For the first, suppose T_1 and T_2 preserve the bracket, and calculate

$$\begin{aligned}(T_1 \circ T_2)([\boldsymbol{v}, \boldsymbol{w}]) &= T_1(T_2([\boldsymbol{v}, \boldsymbol{w}])) \\ &= T_1([T_2(\boldsymbol{v}), T_2(\boldsymbol{w})]) \\ &= [T_1(T_2(\boldsymbol{v})), T_1(T_2(\boldsymbol{w}))] \\ &= [(T_1 \circ T_2)(\boldsymbol{v}), (T_1 \circ T_2)(\boldsymbol{w})].\end{aligned}$$

For the second, write $\boldsymbol{v}' = T^{-1}(\boldsymbol{v})$ and $\boldsymbol{w}' = T^{-1}(\boldsymbol{w})$. (Note the prime does *not* mean differentiation here — just a different vector.) Observe that $T([\boldsymbol{v}', \boldsymbol{w}']) = [T(\boldsymbol{v}'), T(\boldsymbol{w}')] =$

$[\boldsymbol{v}, \boldsymbol{w}]$ implies that $[\boldsymbol{v}', \boldsymbol{w}'] = T^{-1}([\boldsymbol{v}, \boldsymbol{w}])$. But this last equation says

$$[T^{-1}(\boldsymbol{v}), T^{-1}(\boldsymbol{w})] = T^{-1}([\boldsymbol{v}, \boldsymbol{w}]). \quad \square$$

For an appropriately chosen finite-dimensional Lie algebra L, we will define $G(L, \mathbb{F})$ as a subgroup of Aut(L). When the field $\mathbb{F}$ permits differentiation, we can recover the notion of tangent space, and the original Lie algebra L is in fact the tangent space at the identity of this "built-by-hand" group $G(L, \mathbb{F})$.

The following theorem is key to the construction of the group $G(L, \mathbb{F})$. It is due to the French mathematician Claude Chevalley (1909–1984), who made many seminal contributions to this area of mathematics. He was also a founding member of the legendary *Bourbaki* group of mathematicians. A Lie algebra basis of the kind described in Chevalley's theorem is now called a *Chevalley basis*. Groups constructed using this theorem are called *Chevalley groups*. (The definition of a *simple* Lie algebra is in the next subsection.)

Theorem 6.5.2* (Chevalley) *Assume L is a finite-dimensional simple Lie algebra over the complex numbers. Then L has a basis $\{X_1, X_2, \ldots, X_n\}$ for which all the "structure constants" are integers; that is, $[X_i, X_j] = \sum a_{ij\,k} X_k$ and the $a_{ij\,k}$ are integers, for all $i, j, k = 1, \ldots, n$.*

For any field $\mathbb{F}$, we can interpret an integer as an element of $\mathbb{F}$. With this identification, Chevalley's theorem tells us that we can regard L as a Lie algebra over $\mathbb{F}$, as well as over $\mathbb{C}$—just by restricting scalars appropriately.

Return now to working over $\mathbb{C}$. Recall that we can define the adjoint map in terms of the bracket operation: if $X \in L$, then $\mathrm{ad}(X) : L \to L$ by

$$\mathrm{ad}(X)(Y) = [X, Y].$$

By the properties of the bracket, ad(X) is a linear transformation. If X is an element of a Chevalley basis for L, the matrix of ad(X) with respect to this basis has integer entries.

It turns out that for certain of the elements of the Chevalley basis, the matrix of ad(X) is diagonal. Otherwise, it is either *strictly upper triangular* (with zeroes on and below the main diagonal) or *strictly lower triangular* (with zeroes on and above the main diagonal). Let $\mathcal{B}$ be the set of basis elements X giving one of these strictly upper or lower triangular matrices for ad(X), and identify ad(X) with its matrix with respect to the Chevalley basis.

For any strictly upper (or lower) triangular square matrix, it can be shown that some power of the matrix is the zero matrix. As a consequence, if $X \in \mathcal{B}$, then some power of ad(X) is the zero matrix. For these special Xs, the entries in exp(t ad(X)) are polynomials in t, rather than infinite series in t. The coefficients of these polynomials are rational numbers, with integers in the numerators coming from the "structure constants" for the Chevalley basis and integers in the denominators from the coefficients $t^m/m!$ in the terms of the matrix exponential. Because of special properties of the structure constants, it can be proved that the entries in the matrix exp(t ad(X)) are actually polynomials in t with *integer* coefficients.

As we've already observed, any integer can be interpreted as an element of $\mathbb{F}$. Therefore the entries in exp(t ad (X)) for $t \in \mathbb{F}$ are meaningful for any field $\mathbb{F}$. In fact, it turns out that the matrix exp(t ad (X)) even has determinant 1.

Since ad(X) is the matrix of a linear transformation of L (with respect to the Chevalley basis of L),

$$M(t, X) = \exp(t\ \mathrm{ad}(X))$$

is also the matrix of a linear transformation of L.

Example 6.5.3 Let L be the Lie algebra over $\mathbb{C}$ consisting of 3 by 3 matrices A satisfying $A + A^T = 0_3$. As you will see, it really matters that we work over the complex numbers. We know

$$L = \left\{ \begin{bmatrix} 0 & u & v \\ -u & 0 & w \\ -v & -w & 0 \end{bmatrix} : u, v, w \in \mathbb{C} \right\}.$$

A Chevalley basis for L consists of

$$E = \begin{bmatrix} 0 & 0 & 1 \\ 0 & 0 & -i \\ -1 & i & 0 \end{bmatrix}, \quad H = \begin{bmatrix} 0 & i & 0 \\ -i & 0 & 0 \\ 0 & 0 & 0 \end{bmatrix}, \quad F = \begin{bmatrix} 0 & 0 & 1 \\ 0 & 0 & i \\ -1 & -i & 0 \end{bmatrix}.$$

With respect to this basis, the adjoints of the basis elements are

$$\mathrm{ad}(E) = \begin{bmatrix} 0 & -1 & 0 \\ 0 & 0 & -2 \\ 0 & 0 & 0 \end{bmatrix}, \quad \mathrm{ad}(H) = \begin{bmatrix} 1 & 0 & 0 \\ 0 & 0 & 0 \\ 0 & 0 & -1 \end{bmatrix}, \quad \mathrm{ad}(F) = \begin{bmatrix} 0 & 0 & 0 \\ 2 & 0 & 0 \\ 0 & 1 & 0 \end{bmatrix}.$$

How was this basis found? Suppose instead we had started with the natural basis for L:

$$X = \begin{bmatrix} 0 & 1 & 0 \\ -1 & 0 & 0 \\ 0 & 0 & 0 \end{bmatrix}, \quad Y = \begin{bmatrix} 0 & 0 & 1 \\ 0 & 0 & 0 \\ -1 & 0 & 0 \end{bmatrix}, \quad Z = \begin{bmatrix} 0 & 0 & 0 \\ 0 & 0 & 1 \\ 0 & -1 & 0 \end{bmatrix}.$$

With respect to this basis, the matrix of ad(X) is

$$\mathcal{X} = \mathrm{ad}(X) = \begin{bmatrix} 0 & 0 & 0 \\ 0 & 0 & 1 \\ 0 & -1 & 0 \end{bmatrix}.$$

Suppose we choose X to be the vector "H"; in other words, we want to diagonalize $\mathcal{X}$ =ad(X). In the usual way, we find that $\det(\mathcal{X} - \lambda I_3) = -\lambda(\lambda^2 + 1)$, so the eigenvalues of $\mathcal{X}$ are $0, i, -i$. The corresponding eigenvectors turn out to be X, $Y + iZ$ and $Y - iZ$. *But* the Chevalley basis requires *integer* structure constants, and so *integer* eigenvalues are necessary. Therefore, multiply by i and choose $H = iX$, which gives $\mathrm{ad}(H) = i\mathcal{X}$. The eigenvalues of $i\mathcal{X}$ are then $0, 1, -1$, with corresponding eigenvectors H, $E = Y - iZ$ and $F = Y + iZ$. □

Proposition 6.5.4 *The linear transformation* $M(t, X) = \exp(t\ ad(x))$ *is invertible. For* $X \in \mathcal{B}$, $M(t, X)$ *also preserves the bracket operation on* L, *so* $M(t, X)$ *is in* $Aut(L)$.

Proof sketch In matrix notation, invertibility follows since $M(t, X)M(-t, X) = I_n$. Proving that $M(t, X)$ preserves the bracket is more complicated. It depends on the facts that $(\text{ad}(X))^m = 0$ for some m and also that $\delta = \text{ad}(X)$ is a *derivation* of L: a linear transformation $L \to L$ satisfying $\delta([U, V]) = [U, \delta(V)] + [\delta(U), V]$ for all $U, V \in L$. (Notice the resemblance to the product rule for derivatives, which accounts for the name.) This second fact is a consequence of the Jacobi identity for the Lie algebra L. □

Proposition 6.5.4 tells us that $M(t, X)$ is a symmetry of the Lie algebra L—a Lie algebra isomorphism. We define a *group* of symmetries of L in terms of the Lie algebra L (and an associated Chevalley basis) and the field $\mathbb{F}$ by

$$G(L, \mathbb{F}) = \langle\ M(t, X) = \exp(t\ \text{ad}(X)) : t \in \mathbb{F}, X \in \mathcal{B}\ \rangle.$$

What we mean by the angle brackets (in matrix language) is that $G(L, \mathbb{F})$ consists of all matrices that can be written as products of finitely many matrices of the form $M(t, X)$ for $t \in \mathbb{F}$ and $X \in \mathcal{B}$; in other words, $G(L, \mathbb{F})$ is *generated* by the $M(t, X)$ using matrix multiplication. The following proposition tells us that $G(L, \mathbb{F})$ really is a group. Since we already know that the generators $M(t, X)$ are Lie algebra automorphisms of L, it will follow that $G(L, \mathbb{F})$ is a subgroup of Aut(L), as we claimed at the start.

Proposition 6.5.5 *The set of linear transformations $G(L, \mathbb{F})$ is a group under composition of functions. (Equivalently, the set of matrices $G(L, \mathbb{F})$ is a group under matrix multiplication; it is a subgroup of $GL(n, \mathbb{F})$, where n =dim L.)*

The group $G(L, \mathbb{F})$ is called the *Chevalley group of adjoint type* associated with L over $\mathbb{F}$. (When $\mathbb{F}$ is the original field $\mathbb{C}$, we just write $G(L)$.) Obviously the definition of $G(L, \mathbb{F})$ depends on a choice of a Chevalley basis for L. Fortunately, different choices of bases for L produce isomorphic groups, so $G(L, \mathbb{F})$ is determined up to isomorphism by L and $\mathbb{F}$. In many cases, the Chevalley group of adjoint type is a *simple group*. (For readers who have studied abstract algebra: a group is simple if it has no normal subgroups other than itself and the identity subgroup.) Chevalley's construction not only gave new proofs of the simplicity of known families of finite simple groups, but also produced new families. His seminal paper was titled "Sur certains groupes simples," and it appeared in 1955. The French title is no surprise, but less predictable is that the paper appeared in a Japanese journal, *Tôhoku Mathematics Journal.*

Sometimes it turns out that $X \in \mathcal{B}$ is itself strictly upper or lower triangular, so that the entries of $\exp(tX)$ are polynomials in t and can be defined over a general field $\mathbb{F}$. In this case, we have $\text{Ad}(\exp(tX)) = \exp(\text{ad}(tX))$, and we exploit this in Example 6.5.7. Before we turn to the example, we need a fact about $SL(2, \mathbb{F})$, contained in the following proposition.

Proposition 6.5.6 *For any field $\mathbb{F}$, every element of $SL(2, \mathbb{F})$ can be written as a product of matrices of the form* $\begin{bmatrix} 1 & t \\ 0 & 1 \end{bmatrix}$ *and* $\begin{bmatrix} 1 & 0 \\ t & 1 \end{bmatrix}$ *for various choices of $t \in \mathbb{F}$.*

Proof Let $\begin{bmatrix} a & b \\ c & d \end{bmatrix}$ be in $SL(2, \mathbb{F})$, so a, b, c, d are elements of $\mathbb{F}$, and $ad - bc = 1$ in the arithmetic of $\mathbb{F}$.

Case 1: $c \neq 0$.

$$\begin{bmatrix} a & b \\ c & d \end{bmatrix} = \begin{bmatrix} 1 & (a-1)/c \\ 0 & 1 \end{bmatrix} \begin{bmatrix} 1 & 0 \\ c & 1 \end{bmatrix} \begin{bmatrix} 1 & (d-1)/c \\ 0 & 1 \end{bmatrix}.$$

Case 2: $b \neq 0$.

$$\begin{bmatrix} a & b \\ c & d \end{bmatrix} = \begin{bmatrix} 1 & 0 \\ (d-1)/b & 1 \end{bmatrix} \begin{bmatrix} 1 & b \\ 0 & 1 \end{bmatrix} \begin{bmatrix} 1 & 0 \\ (a-1)/b & 1 \end{bmatrix}.$$

Case 3: $b = c = 0$.

$$\begin{bmatrix} a & 0 \\ 0 & 1/a \end{bmatrix} = \begin{bmatrix} 1 & 0 \\ (1-a)/a & 1 \end{bmatrix} \begin{bmatrix} 1 & 1 \\ 0 & 1 \end{bmatrix} \begin{bmatrix} 1 & 0 \\ a-1 & 1 \end{bmatrix} \begin{bmatrix} 1 & -1/a \\ 0 & 1 \end{bmatrix}.$$

This completes the proof. □

Example 6.5.7 We start with

$$L = \{A \in \mathcal{M}(2, \mathbb{C}) : \mathrm{tr}(A) = 0\}.$$

In Chapter 5 you worked with a basis of L that we can now recognize as a Chevalley basis, namely $\{E, H, F\}$, where

$$E = \begin{bmatrix} 0 & 1 \\ 0 & 0 \end{bmatrix}, \quad H = \begin{bmatrix} 1 & 0 \\ 0 & -1 \end{bmatrix}, \quad F = \begin{bmatrix} 0 & 0 \\ 1 & 0 \end{bmatrix}.$$

The basis elements giving strictly upper or lower triangular matrices are $\mathcal{B} = \{E, F\}$. In this case E and F are themselves strictly upper and lower triangular respectively, and we have

$$\exp(tE) = \begin{bmatrix} 1 & t \\ 0 & 1 \end{bmatrix} \quad \exp(tF) = \begin{bmatrix} 1 & 0 \\ t & 1 \end{bmatrix}.$$

We can thus identify $\mathrm{Ad}(\exp(tE))$ with $\exp(t\ \mathrm{ad}(E))$ and think of our group as

$$G(L, \mathbb{F}) = \langle \mathrm{Ad}(\exp(tX)) : X = E \text{ or } X = F, t \in \mathbb{F} \rangle.$$

By Proposition 6.5.6, we can recognize that every matrix M in $SL(2, \mathbb{F})$ appears as $\mathrm{Ad}(M)$ in $G(L, \mathbb{F})$.

However, $G(L, \mathbb{F})$ isn't quite isomorphic to $SL(2, \mathbb{F})$. The two transformations $\mathrm{Ad}(M)$ and $\mathrm{Ad}(-M)$ have the same effect on L, so as elements of $G(L, \mathbb{F})$ they count as the same (much as 2 and 12 count as the same element of $\mathbb{F}_5$). This identification corresponds to what readers who have studied abstract algebra will recognize as working with the "quotient group" $SL(2, \mathbb{F})/\{\pm I_2\}$ instead of with $SL(2, \mathbb{F})$. The two matrices I_2 and $-I_2$ are exactly the elements of $SL(2, \mathbb{F})$ that commute with all the others (so they form a normal subgroup called the *center* of $SL(2, \mathbb{F})$). This quotient group, $G(L, \mathbb{F}) = SL(2, \mathbb{F})/\{\pm I_2\}$, is usually denoted by $PSL(2, \mathbb{F})$. In this notation, 'P' is for "projective" because of a relationship with projective geometry. For the field $\mathbb{F}_2$, we have $-1 = +1$ and $SL(2, \mathbb{F}_2) = PSL(2, \mathbb{F}_2)$, so in this case we have indeed come back to one of our original matrix groups.

All of the groups $PSL(2, \mathbb{F})$ are of interest, and we shall see an important reason why when we take a look at the classification of finite simple groups in the next subsection.

6.5.2 Classifying finite simple groups

A finite group can be made "more simple" by mapping it via a group homomorphism onto a smaller group. When that happens, the group elements that map to the identity of the smaller group constitute a *normal subgroup* of the original group—that is, a subgroup that contains the conjugates of its elements. So, if a group has no normal subgroups (other than itself and the subgroup consisting of the identity alone), it is said to be *simple*, since it can't be mapped by a homomorphism onto a smaller nontrivial group. The finite simple groups are in some sense the building blocks for all finite groups—there is a rough analogy to the way in which the primes are the building blocks (multiplicatively) for the positive integers. Finding finite simple groups has been of interest to mathematicians since the nineteenth century. For example, we have the following theorem.

Theorem 6.5.8* (Dickson, 1901) *$PSL(2, \mathbb{F})$ is simple for every field $\mathbb{F}$ except $\mathbb{F}_2$ and $\mathbb{F}_3$.*

In fact, Chevalley's construction typically leads to a simple group if it begins with a *simple* Lie algebra. The analogue of a normal subgroup of a group is an ideal of a Lie algebra. An *ideal* of a Lie algebra L is a subspace S of L with the following property: if $\boldsymbol{v} \in L$ and $\boldsymbol{u} \in S$, then $[\boldsymbol{v}, \boldsymbol{u}] \in S$. A Lie algebra is *simple* if it contains no ideals other than the subspace consisting of zero alone and L itself.

The simple Lie algebras over $\mathbb{C}$ were classified by W. Killing in papers in 1888–90 and by E. Cartan in his 1894 thesis. They fall into four infinite families of types A, B, C, D plus five exceptional types (see section 5.4). More specifically, the types are denoted A_ℓ for $\ell = 1, 2, \ldots$, B_ℓ for $\ell \geq 2$, C_ℓ for $\ell \geq 3$, and D_ℓ for $\ell \geq 4$. The exceptional types are denoted G_2, F_4 and E_ℓ for $\ell = 6, 7, 8$. Each type of simple Lie algebra is associated with a *Dynkin diagram* comprised of ℓ nodes and either single, double or triple edges between nodes. (See Figure 6.3 below.) One of the manifestations of the unity of mathematics is the fact that Dynkin diagrams describe important phenomena in so many different areas within the discipline.

Remark The cover article of the October 2007 issue of the *Notices* of the American Mathematical Society was about a group G associated with E_8. Here's the opening sentence of the article (by David Vogan of MIT): "On January 8, 2007, just before 9 in the morning, a computer finished writing to disk about sixty gigabytes of files containing the Kazhdan-Lusztig polynomials for the split real group G of type E_8." Mathematicians are still studying groups of Lie type and finding out more.

We can use the example of the (simple) Lie algebra L of the trace zero 2 by 2 matrices over $\mathbb{C}$ to get an idea of where Dynkin diagrams come from in this setting. A Dynkin diagram is associated with a Chevalley basis for the Lie algebra. We know that $\{E, H, F\}$ is a Chevalley basis for L. As we saw in Example (6.5.7), the basis vectors E, H and F are eigenvectors for ad(H), with eigenvalues $+2$, 0 and -2 respectively. The two vectors E and F are paired in the sense that their eigenvalues are negatives of each other. Sometimes you see the notation $\boldsymbol{e}_r = E$, $\boldsymbol{h}_r = H$ and $\boldsymbol{e}_{-r} = F$, where r denotes a function taking H to 2, with $-r$ taking H to -2. The function r is called a *root*. The nodes in the Dynkin diagram are associated with the "simple" roots corresponding

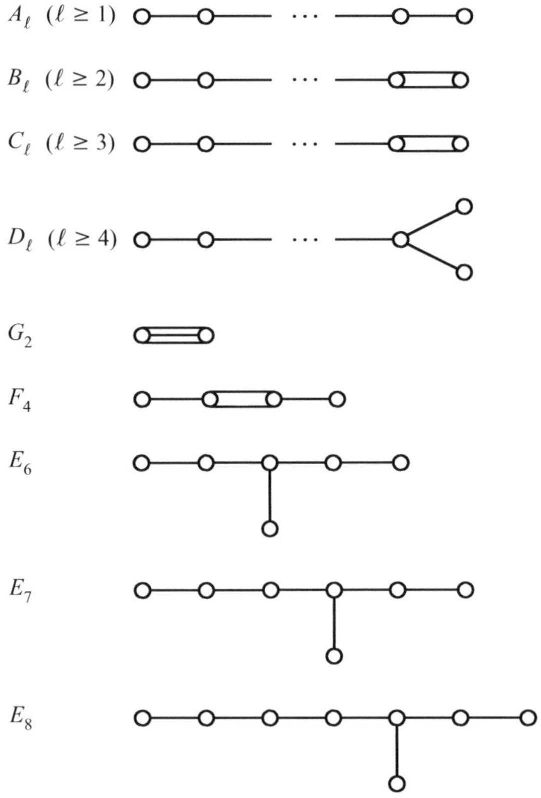

Figure 6.3. Dynkin diagrams associated with simple Lie algebras

to the Chevalley basis. Since there is just one simple root in this case, the Lie algebra corresponds to the diagram for A_1.

More generally,

$$L = \{A \in \mathcal{M}(\ell + 1, \mathbb{C}) : \mathrm{tr}(A) = 0\}$$

is a simple Lie algebra of type A_ℓ. The various $\boldsymbol{h}_r$s are diagonal matrices $\boldsymbol{h}$ with diagonal entries $\{a_0, a_1, \ldots, a_\ell\}$ and with $\mathrm{tr}(\boldsymbol{h}) = \sum_j a_j = 0$. The other basis vectors are the matrices $\boldsymbol{e}_{ij}$ for $i \neq j$ having all entries 0 except the i, j entry equals 1. A calculation shows that $\mathrm{ad}(\boldsymbol{h})(\boldsymbol{e}_{ij}) = (a_i - a_j)\boldsymbol{e}_{ij}$. The "root" associated with $\boldsymbol{e}_{ij}$ takes $\boldsymbol{h}$ to $a_i - a_j$. There are ℓ "simple roots" $r_1, \ldots, r_\ell$, and they have the form $r_j(\boldsymbol{h}) = a_{j-1} - a_j$. The ℓ nodes in the Dynkin diagram for L correspond to the ℓ simple roots. (The explanation of the edges between nodes is more complicated.)

The work of Chevalley in the late 1940s and early 1950s (especially his 1955 paper) inspired a renewed effort to classify the finite simple groups as the simple Lie algebras (and Lie groups) had been classified. Much work on finite groups of matrices was done by Jordan in the 1870s and Dickson around the turn of the 20th century, but then their work fell a bit out of fashion. A major step was a paper by Steinberg in 1959 called, appropriately enough, *Variations on a Theme of Chevalley*. As the finite simple groups "of Lie type" became better understood, hope grew that the classification of all the finite simple groups was near. The finite simple groups seemed to fall into infinite families, one the "alternating groups" and all the others corresponding to the families of simple Lie algebras (including the five exceptional Lie algebras and also some twisted variants analogous to the twist on the family A_ℓ that gives the unitary groups), except for five anomalous "sporadic" groups discovered by Mathieu between 1861 and 1873. However, in 1964 something remarkable happened: a Croatian mathematician named Zvonimir Janko working in Australia discovered a new "sporadic" simple group. Twenty more new sporadic groups were discovered in the next decade. The classification project became much more complicated, and in a stunning collaborative effort involving about 100 mathematicians working for 30 years and producing roughly 500 journal articles totalling approximately 10,000 pages, the classification was completed around 1980. (In fact, a gap in the argument has been filled more recently. And mathematicians are still working to simplify the gigantic proof.)

APPENDIX I

Linear algebra facts

Here are some basic facts from linear algebra that you should know. To start, we'll assume that all our vector spaces are over the real numbers $\mathbb{R}$. We'll use capital letters like A for matrices, lower case letters like a for real numbers, and bold-face lower case letters like $\boldsymbol{a}$ for vectors.

- You should know how to add and scalar multiply vectors (algebraically and, in $\mathbb{R}^2$ or $\mathbb{R}^3$, geometrically). In $\mathbb{R}^2$ you should also know how to interpret an ordered pair of real numbers (a, b) as a *point* or a *vector*; similarly for ordered triples in $\mathbb{R}^3$.
- You should know how to calculate the *dot product* of two vectors in $\mathbb{R}^n$, and you should know the geometric interpretation of the dot product in $\mathbb{R}^2$ or $\mathbb{R}^3$.
- You should know how to calculate the *cross product* of two vectors in $\mathbb{R}^3$, and you should know the geometric interpretation of the cross product. A useful fact about cross products is $\boldsymbol{a} \times (\boldsymbol{b} \times \boldsymbol{c}) = (\boldsymbol{a} \cdot \boldsymbol{c})\, \boldsymbol{b} - (\boldsymbol{a} \cdot \boldsymbol{b})\, \boldsymbol{c}$.
- You should know how to add and multiply matrices of the appropriate sizes and how to multiply a matrix by a scalar; you should know the properties of addition, multiplication and scalar multiplication (e.g., $A(B+C) = AB + AC$). You should know what the n by n identity matrix I_n is and what the n by m zero matrix is, for any positive integers m and n. (Normally, AB denotes matrix multiplication A times B for matrices of appropriate sizes.)
- You should know what the *transpose* A^T of a matrix A is, and that $(rA + sB)^T = (rA)^T + (sB)^T = rA^T + sB^T$ and $(AB)^T = B^T A^T$ for matrices A and B of the appropriate size. Note that if $\boldsymbol{v}, \boldsymbol{w} \in \mathbb{R}^n$ are column vectors (thought of as n by 1 matrices), then $\boldsymbol{v}^T$ and $\boldsymbol{w}^T$ are row vectors, and $\boldsymbol{v} \cdot \boldsymbol{w} = \boldsymbol{v}^T \boldsymbol{w} = \boldsymbol{v}^T I_n\, \boldsymbol{w}$.
- You should know how to calculate the *determinant* of a 2 by 2 or 3 by 3 matrix; you should know that the determinant of an n by n matrix is also defined for all positive integers n, although I don't expect that you know the general definition.
- You should know that $\det(AB) = \det(A)\det(B)$ and $\det(A^T) = \det A$ for square matrices A and B, and that if A' is obtained by multiplying a row (or column) of A by $c \in \mathbb{R}$, then $\det A' = c \det A$.

- You should know that the *trace* of a matrix is the sum of its diagonal entries, and that $\text{tr}(AB) = \text{tr}(BA)$, $\text{tr}(PAP^{-1}) = \text{tr}(A)$, $\text{tr}(A + B) = \text{tr}(A) + \text{tr}(B)$, and $\text{tr}(rA) = r\ \text{tr}(A)$ for $r \in \mathbb{R}$ and for n by n matrices A, B, P with P invertible.
- For vector spaces V and W, you should know the definition of a *linear transformation* $T: V \to W$:
  $T(\boldsymbol{v}_1 + \boldsymbol{v}_2) = T(\boldsymbol{v}_1) + T(\boldsymbol{v}_2)$ and $T(a\boldsymbol{v}) = aT(\boldsymbol{v})$ for $\boldsymbol{v}_1, \boldsymbol{v}_2, \boldsymbol{v} \in V$ and $a \in \mathbb{R}$.
- Given a linear transformation $T: \mathbb{R}^n \to \mathbb{R}^m$, you should know how to find the matrix A of T with respect to specified bases of $\mathbb{R}^n$ and $\mathbb{R}^m$. We'll use the calculus convention, writing functions on the left, so if $\boldsymbol{v}$ is a column vector written with respect to the same basis of $\mathbb{R}^n$, then $T(\boldsymbol{v}) = A\boldsymbol{v}$, a column vector with respect to the specified basis of $\mathbb{R}^m$. In particular, the columns of A are indexed by the basis vectors of the domain $\mathbb{R}^n$, and the rows of A are indexed by the basis vectors of the range $\mathbb{R}^m$.
- You should know that if linear transformations T and S have matrices A and B respectively, then $T \circ S$ has matrix AB. That is, $(T \circ S)(\boldsymbol{v}) = (AB)\boldsymbol{v} = AB\boldsymbol{v}$, assuming the composition $T \circ S$ is defined and that A and B are matrices with respect to corresponding bases. Often we'll write $T \circ S$ as TS.
- You should know a linear transformation $T: \mathbb{R}^n \to \mathbb{R}^n$ is invertible $\Leftrightarrow$ its corresponding n by n matrix A is invertible $\Leftrightarrow$ the determinant of A is nonzero.
- You should know that if A is the matrix of an invertible linear transformation $T: \mathbb{R}^n \to \mathbb{R}^n$ with respect to specified bases, then A^{-1} is the matrix of T^{-1} with respect to the same bases and $A^{-1}A = AA^{-1} = I_n$. Also $\det(I_n) = 1$ and $\det(A^{-1}) = 1/\det(A)$.
- You should know how to find the nullspace of an n by m matrix A or, equivalently, the kernel of the linear transformation $T: \mathbb{R}^m \to \mathbb{R}^n$ with matrix A. Also, you should know that the n by n matrix A is invertible $\Leftrightarrow$ the nullspace of A is $\{\mathbf{0}\}$; equivalently, the linear transformation $T: \mathbb{R}^n \to \mathbb{R}^n$ is invertible $\Leftrightarrow$ the kernel of T is $\{\mathbf{0}\}$.
- You should know that for an invertible 2 by 2 matrix A with entries a_{ij},
$$A^{-1} = \frac{1}{\det(A)} \begin{bmatrix} a_{22} & -a_{12} \\ -a_{21} & a_{11} \end{bmatrix}.$$
- You should know that an *eigenvector* $\boldsymbol{v}$ of a linear transformation $T: V \to V$ is a nonzero vector $\boldsymbol{v} \in V$ satisfying $T(\boldsymbol{v}) = \lambda\boldsymbol{v}$ for some scalar λ called the *eigenvalue* corresponding to $\boldsymbol{v}$. Similarly, if A is a square matrix, a nonzero vector $\boldsymbol{v}$ is an eigenvector for A if $A\boldsymbol{v} = \lambda\boldsymbol{v}$ for some scalar λ.
- You should know λ is an eigenvalue for an n by n matrix $A \Leftrightarrow \det(A - \lambda I_n) = 0$. The corresponding *eigenvector* is found by determining the null space of $A - \lambda I_n$.
- You should know that if A and B are matrices for a linear transformation $T: \mathbb{R}^n \to \mathbb{R}^n$ with respect to different bases of $\mathbb{R}^n$, then there is an invertible matrix P for which $B = PAP^{-1}$. In particular, A and B have the same determinant and trace, which we regard as the determinant and trace, respectively, of T. Also, A, B and T have the same eigenvalues.

- You should *assume* the following. The *characteristic polynomial* of an n by n matrix A is $f(x) = \det(A - xI_n)$. The polynomial $f(x)$ has degree n, the coefficient of x^n is $(-1)^n$, the coefficient of x^{n-1} is $(-1)^{n-1}\operatorname{tr}(A)$, and the constant term is $\det(A)$. Moreover (Cayley-Hamilton Theorem) the matrix A satisfies its *characteristic equation* $f(x) = 0$. That is, $f(A)$ is the zero matrix.

APPENDIX II

Paper assignment used at Mount Holyoke College

Choose one of the topics in the following list. (If there is something not on the list that you would particularly like to explore, you can come talk to me about it.) You may use any books or articles as resources in preparing your paper, but all sources should be acknowledged. You may discuss your work with other students or with me (in fact, this is encouraged!), but the writing should be your own.

The paper should be written in narrative form, in complete sentences and should be 4–6 typed pages in length. Imagine your audience is another student with the same mathematical background as yours and that you are explaining your ideas to her. You should provide the definitions of any terms not already defined in the text and illustrate new terms with examples. You may cite previous results (from your reading or from the text) and use them without proof, as needed. Length considerations will help determine which results you prove in full and which you merely cite. Include at least one proof. Be sure to verify that the necessary hypotheses are satisfied when you apply a theorem.

Whenever possible, relate your topic to the text material. Explain the meaning of any result you prove, and provide enough detail in your argument so that the reasoning would be clear to someone encountering your proof for the first time. Illustrate the results with one or more examples to make the meaning clearer.

- You should review your intended **topic** with me by mid-term so that I can be sure you have the appropriate background to handle it effectively. As a general guideline, if you haven't taken Abstract Algebra, choose from among topics 1 – 6. If you have taken Abstract Algebra but not Real Analysis, choose from among topics 1–13.
- At least 3 weeks before the paper is due, you should give me an **outline** indicating which statements you will prove and what examples you will use, so I can be sure you aren't attempting too much or too little.
- The completed **paper** is due by the last week of classes.

Possible topics:

1. Discuss world lines and the partition of spacetime. Show that the rotations of the first kind map the "future set" one-to-one and onto itself. Reference: Callahan, pp. 1–9 and pp. 43–59.

2. Discuss Howe's claim that many of the important classical differential equations are related to Lie theory, and illustrate by doing some of the exercises he provides. Reference: Howe, p. 622, Exercises A and B.

3. There is a way to associate with a matrix A in $SL(2, \mathbb{R})$ a *Moebius transformation* $m: \mathbb{C} \to \mathbb{C}$. Specifically, the matrix

$$A = \begin{bmatrix} a & b \\ c & d \end{bmatrix} \in SL(2, \mathbb{R})$$

coresponds to the Moebius transformation $m = \Phi(A)$, where

$$m(z) = \frac{az + b}{cz + d}.$$

First show that Φ is a homomorphism; that is, show that if

$$A_j = \begin{bmatrix} a_j & b_j \\ c_j & d_j \end{bmatrix} \in SL(2, \mathbb{R}), j = 1, 2$$

and $\Phi(A_j) = m_j$, then $\Phi(A_1 A_2) = m_1 \circ m_2$. Also show that the Moebius transformations coming from $SL(2, \mathbb{R})$ are isometries of the Poincaré half-plane. More specifically, show that if $w = m(z)$, for $z = x + iy$ with Im $z = y > 0$, then $w = x' + iy'$ with Im $w = y' > 0$. Also the metric on the upper half-plane is preserved by m. That is,

$$\frac{dw\, d\overline{w}}{(\text{Im}\, w)^2} = \frac{dz\, d\overline{z}}{(\text{Im}\, z)^2}$$

where $dw = (dw/dz)dz = [1/(cz + d)^2]\, dz$ and $d\overline{w} = [1/(c\overline{z} + d)^2]\, d\overline{z}$. Reference: Sattinger and Weaver, pp. 8–10. (Also see Mumford, *et al*, pp. 69–77 on Moebius transformations coming from $GL(2, \mathbb{R})$.)

4. The *Riemann sphere* is a unit sphere centered at the origin of $\mathbb{R}^3$ thought of as $\mathbb{C} \cup \{\infty\}$ via stereographic projection. The stereographic projection is described on page 10 of Sattinger and Weaver. Note that they use the Greek letters ξ, η, ζ (called xi, eta and zeta respectively) for coordinates in $\mathbb{R}^3$, reserving x and y for the real and imaginary parts of a complex number. Thus the point $(\xi, \eta, \zeta) \neq (0, 0, 1)$ on the sphere corresponds to the complex number $x + iy$ (and $(0, 0, 1)$ corresponds to ∞). Explain why the equations in Figure 1.33 describing the stereographic projection are correct. Which points on the sphere correspond to complex numbers of length 1? Which correspond to nonzero complex numbers of length less than 1? Which corresponds to the complex number 0? Explain why rotating the sphere around the ζ axis counter-clockwise through an angle of γ corresponds to the map $\mathbb{C} \to \mathbb{C}$ given by

$$z \mapsto \frac{az + b}{cz + d} \quad \text{with } a = e^{i\gamma/2},\ b = 0,\ c = 0,\ d = e^{-i\gamma/2}.$$

What happens to ∞ under this map? (Use the usual conventions for

$$\lim_{t \to \infty} (at + b)/(ct + d)$$

when a, b, c, d are fixed numbers.) Does this correspond to the effect of the rotation on $(0, 0, 1)$? Reference: Sattinger and Weaver, p. 10.

5. Continuing the notation of the previous topic, explain the determination of the 2 by 2 matrices corresponding to the rotations around the ξ and η axes on pages 11 and 12 of Sattinger and Weaver. Verify the claim that the three 2 by 2 matrices in this and the preceding item are exponentials of the *Pauli spin matrices* as described on pages 12 and 13. (Note that this and the preceding topic are the ingredients in Sattinger and Weaver's proof that $SU(2,\mathbb{C})$ is the universal covering group of the Euclidean $SO(3,\mathbb{R})$.) Reference: Sattinger and Weaver, pp. 11–13.

6. The group $SL(2,\mathbb{C})$ is the universal covering group of the Lorentz group $\mathcal{L}^{++}(4,\mathbb{R})$. Give part of the proof that this is so by showing that there is a group homomorphism mapping $SL(2,\mathbb{C})$ to $\mathcal{L}(4,\mathbb{R})$. Specifically (using notation different from the reference), show that the following statements are true.

   (a) The vectors $\boldsymbol{v} = (t, x, y, z)$ of space time can be represented by matrices
   $$V = \begin{bmatrix} t+z & x-iy \\ x+iy & t-z \end{bmatrix}.$$
   Using this representation, the Lorentz product $\boldsymbol{v}\cdot\boldsymbol{v} = \det(V)$. Note that the product of any two vectors $\boldsymbol{v}$ and $\boldsymbol{w}$ can be written
   $$\boldsymbol{v}\cdot\boldsymbol{w} = (1/2)[(\boldsymbol{v}+\boldsymbol{w})\cdot(\boldsymbol{v}+\boldsymbol{w}) - \boldsymbol{v}\cdot\boldsymbol{v} - \boldsymbol{w}\cdot\boldsymbol{w}],$$
   and thus $\boldsymbol{v}\cdot\boldsymbol{w}$ can also be expressed using determinants. In particular, $T\colon \mathbb{R}^4 \to \mathbb{R}^4$ preserves the Lorentz dot product in general if $T(\boldsymbol{v})\cdot T(\boldsymbol{v}) = \boldsymbol{v}\cdot\boldsymbol{v}$ for all $\boldsymbol{v}\in\mathbb{R}^4$.

   (b) The coordinates of $\boldsymbol{v}$ can be recovered from the entries of V by $t = (1/2)\mathrm{tr}(V)$, $x = (1/2)\mathrm{tr}(V\sigma_1)$, $y = (1/2)\mathrm{tr}(V\sigma_2)$, $z = (1/2)\mathrm{tr}(V\sigma_3)$, where the σ_j are the Pauli spin matrices:
   $$\sigma_1 = \begin{bmatrix} 0 & 1 \\ 1 & 0 \end{bmatrix}, \quad \sigma_2 = \begin{bmatrix} 0 & -i \\ i & 0 \end{bmatrix}, \quad \sigma_3 = \begin{bmatrix} 1 & 0 \\ 0 & -1 \end{bmatrix}.$$

   (c) For $A \in SL(2,\mathbb{C})$, define $S_A\colon \mathbb{R}^4 \to \mathbb{R}^4$ by $S_A(V) = AV\overline{A}^T$. Show $S_A \in \mathcal{L}(4,\mathbb{R})$. [In fact, S_A is a rotation of the first kind, so the map $\Phi(A) = S_A$ is actually from $SL(2,\mathbb{C})$ to $\mathcal{L}^{++}(4,\mathbb{R})$.]

   (d) Show that $\Phi(A) = S_A$ defines a *group homomorphism*.

   Reference: Sattinger and Weaver, pp. 16–18.

7. Describe the *intrinsic distance* on the 3-sphere S^3, and show that conjugation by elements of the group S^3 is an isometry preserving this distance. Reference: Zulli, pp. 223–225. Also see Conway and Smith pp. 23–24.

8. Describe the relationship between longitudes of the 3-sphere and conjugacy classes of the group S^3. Reference: Zulli, pp. 225-227, or Artin, pp. 273–276.

9. Prove that the group of inner automorphisms of the quaternions is isomorphic to the (Euclidean) group $SO(3,\mathbb{R})$. Reference: Burn, chapter 19, pp. 179–183. (Note that *codomain* means the same thing as *range*, and *bijection* means the same thing as *one-to-one and onto function*.) Also see Conway and Smith, pp. 23–24.

10. Prove that the normalizer of a Lie subalgebra is a Lie subalgebra. Also prove that the centralizer of a Lie subalgebra is a Lie subalgebra. Reference: Humphreys, pp. 6–7.

11. Prove that the derivations $Der(L)$ of a Lie algebra L form a Lie algebra and that the inner derivations form an ideal of $Der(L)$. Reference: Humphreys, pp. 4, 6.

12. Suppose that $\mathbb{F}$ is a finite field containing q elements. (If q is a prime, you may think of $\mathbb{F}$ as the set $\{0, 1, 2, \ldots, q-1\}$ with addition and multiplication defined modulo q.) Show that $GL(2, q)$, the group of all invertible 2 by 2 matrices with entries in $\mathbb{F}$, contains $(q^2 - 1)(q^2 - q)$ matrices. Also show that $SL(2, q)$, the subgroup of $GL(2, q)$ consisting of matrices of determinant 1, contains $(q^2 - 1)(q^2 - q)/(q - 1) = (q^2 - 1)q$ elements. Reference: Gorenstein, p. 40.

13. Investigate the Lie algebra isomorphisms among the classical Lie algebras of small rank. In particular, working over $\mathbb{R}$ (or over an arbitrary field), show that the classical Lie algebras of type A_1, B_1 and C_1 are all isomorphic. Show that the Lie algebra of type D_1 is 1-dimensional. Show that the Lie algebras of type B_2 and C_2 are isomorphic, as are the Lie algebras of type D_3 and A_3. What can you say about the Lie algebra of type D_2? Reference: Humphreys, pp. 1–6 (see exercise 10 on p. 6).

14. Prove that for a connected Lie group G, if U is a neighborhood of the identity, then every element of G is a finite product of elements of U. Reference: Hausner and Schwartz, Lemma 2, p. 37.

15. This is an alternate approach to some of the ideas in the text. For a matrix group $G \subseteq GL(n, \mathbb{R})$, define

$$\mathcal{G} = \{A \in M(n, \mathbb{R}) : \exp(tA) \in G \text{ for all } t \in \mathbb{R}\}.$$

Prove that

a) $\mathcal{G}$ is a Lie algebra, and

b) $\exp\colon \mathcal{G} \to G$ maps a neighborhood of 0 in $\mathcal{G}$ one-to-one and onto a neighborhood of the identity in G.

Reference: Howe, Theorem 17 and Lemma 18, pp. 616–618.

APPENDIX III

Opportunities for further study

Here are some possibilities for study and investigation to extend and build upon the material of this text. This appendix expands upon some of the suggestions for further reading in *Putting the pieces together* and some ideas from the *Broader view* sections at the end of each chapter.

Metric vector spaces and symmetries

Although Snapper and Troyer say that their *Metric Affine Geometry* requires an introduction to modern algebra that includes normal subgroups and quotient groups, much of the material in their chapter 2 on Metric Vector Spaces is accessible to any student who has worked through the problems in this book. The special case of the quotient of a vector space by a subspace is reviewed in detail when it first appears. Snapper and Troyer classify metric vector spaces over fields of characteristic different from 2 (i.e., fields in which the integer 2 is interpreted as a nonzero element, so "division by 2" is possible), including Sylvester's theorem. But they do more, also proving two "deep and beautiful" structure theorems: Witt's theorem and the Cartan–Dieudonné theorem. Their treatment of the commutator subgroup really does require more background from abstract algebra, but the exposition of Witt's theorem is accessible without it. There are many exercises in the book with which students can test and expand their understanding. A student with more background from abstract algebra might even tackle Emil Artin's classic *Geometric Algebra*. As Snapper and Troyer write in their preface, "We hope that anyone who has worked through our second chapter can read *Geometric Algebra* from the geometric viewpoint which Artin found so important." (By the way, Emil Artin was the father of Michael Artin, whose *Algebra* has been referred to several times in this book.)

Another approach is in Paul Yale's *Geometry and Symmetry*. This book presents the tools from abstract algebra that it requires, so the prerequisites can be flexible, although a student who has studied abstract algebra could move more rapidly and cover more. Unlike Snapper and Troyer, Yale includes discrete and finite groups of isometries and covers the classification of the crystallographic point and space groups. (Yale, like Snapper and Troyer, also includes projective geometry.)

Lie algebras and Chevalley groups

Here we give two suggestions for a student with a good background in abstract algebra (including quotient structures in groups, rings and vector spaces).

For a focus on Lie algebras as algebraic systems, working through the first four chapters of J.E. Humphreys' *Introduction to Lie Algebras and Representation Theory* would make a challenging semester. The first chapter, containing definitions and examples and culminating in Engel's theorem on nilpotent Lie algebras, is particularly accessible. Each chapter includes examples and exercises. The exposition is terse, but very clear. Humphreys doesn't get to the Chevalley basis and Chevalley groups until chapter VII.

Another path, reaching the Chevalley groups more quickly, would be based on R.W. Carter's *Simple Groups of Lie Type*. Carter achieves this by keeping the discussion of Lie algebras to a minimum and omitting any representation theory. The first four chapters cover the classical simple groups, Weyl groups, simple Lie algebras and Chevalley groups. There are unfortunately no exercises in Carter's book, but the exposition is clear, and filling in details of arguments and examples would solidify the ideas. An ambitious student might go farther, at least sampling from chapters 5–8 on the structure of finite Chevalley groups. The last chapter of Carter's book is a brief survey of what was known about the sporadic simple groups at the time of writing, when the classification was not yet complete. (In fact Carter's list of sporadic groups was incomplete as well.)

Quaternions and octonions

Exploring some familiar ideas over the quaternions or the octonions makes a good project that requires only linear algebra as a prerequisite. There are surprises even for 2 by 2 matrices over the quaternions. For example, even though an n by n complex *Hermitian* matrix A (i.e., one satisfying $\overline{A}^T = A$) has n real eigenvalues (counting multiplicity), a 2 by 2 Hermitian matrix with quaternion entries can have eigenvalues which are not real. Perhaps even stranger, if A is an n by n matrix with entries in $\mathbb{H}$ and $\boldsymbol{v} \in \mathbb{H}^n$ is an eigenvector for A, it can happen that $c\boldsymbol{v}$ is *not* an eigenvector for A for some $c \in \mathbb{H}$. Even stranger things can happen over the octonions. The paper "The Octonionic Eigenvalue Problem" by Dray and Manogue has an especially readable introduction on the complex and quaternion cases, and the octonionic results are fairly accessible and afford the student many opportunities to work out specific examples in detail. In fact, not turning too quickly to the paper can make it more of an independent investigation for the student.

There is a wealth of material in Conway and Smith's *On Quaternions and Octonions: Their Geometry, Arithmetic and Symmetry*, enough for several major projects. Their chapters 2–4 cover some of the same ground as this text, but from a different point of view. They include a number of new topics too. There are no exercises, but filling in the details and working out examples would be a challenge as well as a good test of understanding. Their chapter 6 includes a proof of the theorem by Hurwitz referred to in our section 2.5.

Connections to physics

Callahan's *The Geometry of Spacetime* has exactly the same mathematical prerequisites as this text. The only physics it presupposes is, as he writes, "familiarity with the physical applications of calculus." The text is rich in exercises (and also in historical context). Callahan's preface suggests chapters 1–3 and 5–6 as the core of his book.

Singer's *Symmetry in Mechanics* adds to Callahan's prerequisites familiarity with the content of a standard year-long, calculus-based physics course. A student with only these prerequisites would have to start at the beginning and move systematically through the book, including the embedded exercises. Chapters 5 and 6 are on Lie groups and Lie algebras respectively, familiar to a reader of this text but looking different because of the use of differential forms and elements of "symplectic geometry," introduced starting in chapter 2. (This is the book that converted a mathematician I know from aspiring to be a physicist to deciding to be a mathematician.)

Singer's *Linearity, Symmetry, and Prediction in the Hydrogen Atom* is aimed at "senior-level" mathematics, physics and chemistry majors. The core prerequisite is described briefly as "solid understanding of calculus and familiarity with either linear algebra or advanced quantum mechanics." She goes on to say that while an understanding of Fourier theory is not required, since many students will have seen them, Fourier series and transforms come up in the book fairly often, but these portions can be skipped. There are numerous exercises plus suggested paper topics.

Lie groups as manifolds

Students who have studied both abstract algebra and real analysis can go farther and directly engage the analysis and topology undergirding the study of Lie groups, properly defined as special differentiable manifolds.

K. Tapp's *Matrix Groups for Undergraduates* includes a chapter (chapter 7) summarizing the analysis background required, proving the theorem giving the correspondence — via a diffeomorphism — between a neighborhood of the identity in the Lie group and a neighborhood of zero in the Lie algebra, and defining manifolds. With this background, he proves the Campbell-Baker-Hausdorff theorem and uses it to prove the one-to-one correspondence between Lie subalgebras of $\mathcal{M}(n, \mathbb{R})$ and path-connected subgroups of $GL(n, \mathbb{R})$. From there he moves on to a study of maximal tori. Tapp's book does include exercises in every chapter.

For some students, the last three chapters of Tapp's book might not be a full semester's work. A supplement might be based on W. Rossmann's *Lie Groups: An Introduction Through Linear Groups*. Rossman introduces manifolds in chapter 4, which he uses to define integration on manifolds (chapter 5), and then goes on to representation theory (chapter 6), including a proof of the Peter-Weyl theorem for linear groups. Rossman's book contains problems also, but is somewhat more tersely written than Tapp's.

Solutions to selected problems

1. Symmetries of vector spaces

1.1. What is a symmetry?

Problem 1.1.3.

We can choose for S_1 the rotation counter-clockwise through an angle $3\pi/2$ (or, equivalently, clockwise though an angle $\pi/2$).

Problem 1.1.5.

a. The non-square rectangle has two reflection symmetries; the two mirror lines are the x-axis and the y-axis. It has two rotation symmetries; they can be described as rotations through angles of 0 and π counter-clockwise about the origin.

Problem 1.1.7.

a. Geometrically, flipping the square over the x-axis and then flipping it over the x-axis again restores it to its original position. So the reflection is its own inverse.

b. $T_2^{-1} = T_2$ because $T_2(T_2(x, y)) = T_2(x, -y) = (x, -(-y)) = (x, y)$.

c. Geometrically, a point on the mirror line is (equals) its own mirror image. (This is true no matter what the mirror line of the reflection is.) $T_2(x, 0) = (x, 0)$ for each point $(x, 0)$ on the x-axis.

d. There are no fixed points for T_2 off the x-axis, since $T(x, y) = (x, -y) = (x, y)$ implies $y = 0$. This is true in general because if a point is some positive distance from the mirror line, its mirror image is that same positive distance on the other side of the mirror line, so a point not on the mirror line can't equal its mirror image. In other words, such a point can't be fixed by the reflection.

Problem 1.1.8.

Suppose $T_2(a, b) = T_2(c, d)$ for some points (a, b) and (c, d) in $\mathbb{R}^2$. This means $(a, -b) = (c, -d)$. But two points are the same if and only if their coordinates are equal, so $a = c$ and $-b = -d$. In other words, $(a, b) = (c, d)$. Therefore T_2 is one-to-one. Now, choose (a, b) in $\mathbb{R}^2$. We want to find (x, y) in $\mathbb{R}^2$ with $T_2(x, y) = (a, b)$. Choose $x = a$ and $y = -b$: $T_2(a, -b) = (a, b)$. Therefore T_2 is onto.

Problem 1.1.10.

Because T is a linear transformation, $T(\mathbf{0}) = T(\mathbf{0} + \mathbf{0}) = T(\mathbf{0}) + T(\mathbf{0})$. Every vector of $\mathbb{R}^n$ has an additive inverse, so $T(\mathbf{0})$ does. Add $-T(\mathbf{0})$ to both sides of the equality (i.e., subtract $T(\mathbf{0})$ from both sides):

$$\begin{aligned} T(\mathbf{0}) - T(\mathbf{0}) &= T(\mathbf{0}) + T(\mathbf{0}) - T(\mathbf{0}), \\ \mathbf{0} &= T(\mathbf{0}). \end{aligned}$$

1.2. Distance is fundamental

Problem 1.2.5.

The first matrix describes a rotation counter-clockwise about the origin through an angle of α. Its determinant is 1.

Problem 1.2.8.

After using trigonometric identities,

$$\begin{bmatrix} \cos(\alpha) & -\sin(\alpha) \\ \sin(\alpha) & \cos(\alpha) \end{bmatrix} \begin{bmatrix} \cos(\beta) & \sin(\beta) \\ \sin(\beta) & -\cos(\beta) \end{bmatrix} = \begin{bmatrix} \cos(\alpha+\beta) & \sin(\alpha+\beta) \\ \sin(\alpha+\beta) & -\cos(\alpha+\beta) \end{bmatrix}$$

so we see $T \circ S$ is a reflection across a line through the origin making an angle $(\alpha+\beta)/2$ with the positive x-axis. When the order is reversed, the result is again a reflection, but this time the mirror line makes an angle $(\beta-\alpha)/2$ with the positive x-axis. Comparing matrices, the order *does* matter unless $(\alpha+\beta) = (\beta-\alpha) + 2k\pi$ for some integer k; that is, the order matters unless $\alpha = k\pi$ for some integer k. Notice that when k is even, this says the rotation is I_2, and when k is odd, this says the rotation is $-I_2$.

The special case described corresponds to $\alpha = \pi/2$ and $\beta = \pi/2$ and gives $(T \circ S)(1,0) = (-1,0)$ and $(T \circ S)(0,1) = (0,1)$, which is the reflection across the y-axis. (Remember, $T \circ S$ means do S first.) In the other order, $(S \circ T)(1,0) = (1,0)$ and $(S \circ T)(0,1) = (0,-1)$, which is the reflection across the x-axis.

1.3. Groups of symmetries

Problem 1.3.2

a. From (1.2.7), we know a rotation through an angle α times a rotation through an angle β is a rotation through an angle $\alpha+\beta$ (or, equivalently, through $(\alpha+\beta)-2\pi$). Since the product of two rotations is a rotation, the set is closed under multiplication. The identity is a rotation through an angle of 0, so it is in the set. The inverse of a rotation through an angle α is a rotation through an angle $-\alpha$ (or, equivalently, through $2\pi-\alpha$). So the inverse of a rotation is a rotation.

b. From (1.2.9) we know that the product of two reflections is a rotation, so the set of reflections is *not* closed under multiplication. It also fails to include the identity I_2.

Problem 1.3.5

a. $I(\boldsymbol{v}) \cdot I(\boldsymbol{w}) = \boldsymbol{v} \cdot \boldsymbol{w}$, so $I \in O(3, \mathbb{R})$.

b.

$$\begin{aligned} (T \circ S)(\boldsymbol{v}) \cdot (T \circ S)(\boldsymbol{w}) &= T(S(\boldsymbol{v})) \cdot T(S(\boldsymbol{w})) \\ &= S(\boldsymbol{v}) \cdot S(\boldsymbol{w}) \text{ (T preserves the dot product)} \\ &= \boldsymbol{v} \cdot \boldsymbol{w} \text{ (S preserves the dot product).} \end{aligned}$$

c. Write $\boldsymbol{v} = (v_1, v_2, v_3)$. Choose $\boldsymbol{w} = (1, 0, 0)$; then $\boldsymbol{v} \cdot \boldsymbol{w} = v_1 = 0$. Similarly, choose $\boldsymbol{w} = (0, 1, 0)$ to see $v_2 = 0$, and choose $\boldsymbol{w} = (0, 0, 1)$ to see $v_3 = 0$.

d. If $T(\boldsymbol{v}) = (0, 0, 0)$, then $0 = T(\boldsymbol{v}) \cdot T(\boldsymbol{w}) = \boldsymbol{v} \cdot \boldsymbol{w}$ for any choice of $\boldsymbol{w}$. So, by (c), $\boldsymbol{v} = (0, 0, 0)$.

e. Write $T^{-1}(\boldsymbol{v}) = \boldsymbol{v}'$ and $T^{-1}(\boldsymbol{w}) = \boldsymbol{w}'$, so $T(\boldsymbol{v}') = \boldsymbol{v}$ and $T(\boldsymbol{w}') = \boldsymbol{w}$. Now $T^{-1}(\boldsymbol{v}) \cdot T^{-1}(\boldsymbol{w}) = \boldsymbol{v}' \cdot \boldsymbol{w}' = T(\boldsymbol{v}') \cdot T(\boldsymbol{w}')$, since T preserves the dot product. But this says $T^{-1}(\boldsymbol{v}) \cdot T^{-1}(\boldsymbol{w}) = \boldsymbol{v} \cdot \boldsymbol{w}$, as desired.

Problem 1.3.7

a. $I_3^T I_3 = I_3 I_3 = I_3$, so $I_3 \in O(3, \mathbb{R})$.

b. We're given $M_1, M_2 \in O(3, \mathbb{R})$, so we can assume $M_1^T M_1 = M_2^T M_2 = I_3$. Then $(M_1 M_2)^T (M_1 M_2) = (M_2^T M_1^T)(M_1 M_2) = M_2^T (M_1^T M_1) M_2 = M_2^T I_3 M_2 = M_2^T M_2 = I_3$, so $M_1 M_2 \in O(3, \mathbb{R})$.

c. $M^T M = I_3$ tells us $M^{-1} = M^T$ and $I_3 = M M^T = (M^T)^T (M^T)$, so $M^{-1} = M^T \in O(3, \mathbb{R})$.

d. $1 = \det(M^T M) = \det(M^T) \det(M) = \det(M) \det(M)$, so $\det(M)$ is a root of $x^2 = 1$, and $\det(M) = 1$ or -1.

1.4. Bilinear forms and symmetries of spacetime

Problem 1.4.1.

d. If $\boldsymbol{v} = (t, x)$, then $\boldsymbol{v} \cdot (1, 0) = t = 0$ and $\boldsymbol{v} \cdot (0, 1) = -x = 0$ implies $\boldsymbol{v} = (0, 0)$.

Problem 1.4.7.

a. Since $1 = (1, 0) \cdot (1, 0) = (a, b) \cdot (a, b) = a^2 - b^2$ and $a > 0$, (a, b) is on the branch of the hyperbola parameterized by $(a, b) = (\cosh(\alpha), \sinh(\alpha))$ for some $\alpha \in \mathbb{R}$.

b. $0 = (1, 0) \cdot (0, 1) = (a, b) \cdot (c, d) = c \cosh(\alpha) - d \sinh(\alpha)$, so $c = [\sinh(\alpha)/\cosh(\alpha)]d$.

c. $-1 = (0, 1) \cdot (0, 1) = (c, d) \cdot (c, d) = c^2 - d^2 = [\sinh^2(\alpha)/\cosh^2(\alpha) - 1]d^2 = -d^2/\cosh^2(\alpha)$, so $d^2 = \cosh^2(\alpha)$.
(i) If $d = \cosh(\alpha)$ then $c = \sinh(\alpha)$.
(ii) If $d = -\cosh(\alpha)$, then $c = -\sinh(\alpha)$.

d.

$$\text{(i)} \quad \begin{bmatrix} \cosh(\alpha) & \sinh(\alpha) \\ \sinh(\alpha) & \cosh(\alpha) \end{bmatrix}, \qquad \text{(ii)} \quad \begin{bmatrix} \cosh(\alpha) & -\sinh(\alpha) \\ \sinh(\alpha) & -\cosh(\alpha) \end{bmatrix}.$$

In case (i) the determinant is 1, and in case (ii) it is -1.

Problem 1.4.8.

The matrix for $T \circ S$ is

$$\begin{bmatrix} \cosh(\alpha) & \sinh(\alpha) \\ \sinh(\alpha) & \cosh(\alpha) \end{bmatrix} \begin{bmatrix} \cosh(\beta) & \sinh(\beta) \\ \sinh(\beta) & \cosh(\beta) \end{bmatrix} = \begin{bmatrix} \cosh(\alpha+\beta) & \sinh(\alpha+\beta) \\ \sinh(\alpha+\beta) & \cosh(\alpha+\beta) \end{bmatrix},$$

which is the same as the matrix for $S \circ T$.

2. Complex numbers, quaternions and geometry

2.1. Complex numbers

Problem 2.1.6.

a. $|T_\alpha(z)| = |e^{i\alpha}z| = |e^{i\alpha}||z| = |z|$.

b. $T_\alpha(z_1 + z_2) = e^{i\alpha}(z_1 + z_2) = e^{i\alpha}z_1 + e^{i\alpha}z_2 = T_\alpha(z_1) + T_\alpha(z_2)$, and $T_\alpha(rz) = e^{i\alpha}(rz) = r(e^{i\alpha}z) = rT_\alpha(z)$.

c. The basis vectors are 1 and i. We get $T_\alpha(1) = e^{i\alpha} = \cos(\alpha)1 + \sin(\alpha)i$, $T_\alpha(i) = e^{i\alpha}i = -\sin(\alpha)1 + \cos(\alpha)i$, so the matrix of T_α is the rotation matrix

$$\begin{bmatrix} \cos(\alpha) & -\sin(\alpha) \\ \sin(\alpha) & \cos(\alpha) \end{bmatrix}.$$

2.2. Quaternions

Problem 2.2.2.

b. $z\overline{z} = a^2 + b^2 + c^2 + d^2$, and $1/z = \overline{z}/z\overline{z} = \dfrac{a}{a^2+b^2+c^2+d^2} - \dfrac{b}{a^2+b^2+c^2+d^2}i - \dfrac{c}{a^2+b^2+c^2+d^2}j - \dfrac{d}{a^2+b^2+c^2+d^2}k$.

d. We see from the definition of multiplication that the product zz' is in $\mathbb{H}$ if $z, z' \in \mathbb{H}$. If $z, z' \neq 0$, then $z\overline{z} \neq 0$ and $z'\overline{z'} \neq 0$. Then $(zz')\overline{zz'} = z(z'\overline{z'})\overline{z} = (z'\overline{z'})(z\overline{z})$, since real numbers commute with quaternions. Thus $zz'\overline{zz'}$ is the product of two nonzero real numbers and is also nonzero, so $zz' \neq 0$. The multiplicative identity $1 = 1 + 0i + 0j + 0k$ is a nonzero element of $\mathbb{H}$, and if $z \neq 0$, then its multiplicative inverse $1/z$ is also a nonzero element of $\mathbb{H}$.

2.3. The geometry of rotations of $\mathbb{R}^3$

Problem 2.3.1.

a. If the angle of rotation is 0, then every nonzero vector is an eigenvector with eigenvalue 1. If the angle of rotation is π, then every nonzero vector is an eigenvector with eigenvalue -1. For any other angle, there are no eigenvectors.

b. Any nonzero vector lying on the mirror line is an eigenvector with eigenvalue 1. Any nonzero vector perpendicular to the mirror line is an eigenvector with eigenvalue -1. There are no others. The choice of mirror line doesn't affect *whether* there are eigenvectors, but it affects *which* vectors are eigenvectors.

c. Every nonzero vector on the axis (including $\boldsymbol{a}$) is an eigenvector with eigenvalue 1. If the angle of rotation is zero, every nonzero vector is an eigenvector with eigenvalue 1. If the angle of rotation is π, every nonzero vector that is orthogonal to the axis is an eigenvector with eigenvalue -1. For other angles there are no eigenvectors off the axis. As for (b), the choice of axis affects *which* vectors are eigenvectors.

Problem 2.3.4.

d. Since $T: U \to U$ preserves distance, we know the matrix of T has one of the two following forms:

$$\begin{bmatrix} \cos(\alpha) & -\sin(\alpha) & 0 \\ \sin(\alpha) & \cos(\alpha) & 0 \\ 0 & 0 & 1 \end{bmatrix} \quad \text{or} \quad \begin{bmatrix} \cos(\alpha) & \sin(\alpha) & 0 \\ \sin(\alpha) & -\cos(\alpha) & 0 \\ 0 & 0 & 1 \end{bmatrix}.$$

The determinants are 1 and -1 respectively, so only the first can be the matrix of T.

Problem 2.3.6.

a. It's straightforward to check that $M_1^T M_1 = M_2^T M_2 = I_3$.

b. T_1 is a rotation around the z-axis through an angle of α_1 (counter-clockwise when viewed from the positive z-axis). T_2 is a rotation around the y-axis through an angle of α_2 (counter-clockwise when viewed from the negative y-axis).

c. First rotate through an angle α_1 around the z-axis to bring the point into the xz-plane. Then rotate through an angle α_2 round the y-axis to bring the point to the positive z-axis. Since (a, b, c) was at distance 1 from the origin and rotations preserve distance, it will end up at $(0, 0, 1)$.

Problem 2.3.8.

a. Choose

$$M_1 = \begin{bmatrix} \sqrt{2}/2 & \sqrt{2}/2 & 0 \\ -\sqrt{2}/2 & \sqrt{2}/2 & 0 \\ 0 & 0 & 1 \end{bmatrix}, \qquad M_2 = \begin{bmatrix} 1/\sqrt{3} & 0 & -\sqrt{2}/\sqrt{3} \\ 0 & 1 & 0 \\ \sqrt{2}/\sqrt{3} & 0 & 1/\sqrt{3} \end{bmatrix}.$$

From this it follows that the matrix of S is

$$M = M_2 M_1 = \begin{bmatrix} 1/\sqrt{6} & 1/\sqrt{6} & -\sqrt{2}/\sqrt{3} \\ -1/\sqrt{2} & 1/\sqrt{2} & 0 \\ 1/\sqrt{3} & 1/\sqrt{3} & 1/\sqrt{3} \end{bmatrix}.$$

Problem 2.3.11.

a. The matrix of $T \circ S$ is

$$M = \begin{bmatrix} 0 & -1 & 0 \\ 0 & 0 & -1 \\ 1 & 0 & 0 \end{bmatrix}.$$

The null space of $M - I_3$ is spanned by $\boldsymbol{a} = (-1, 1, -1)$, and this is the axis of $T \circ S$.

3. Linearization

3.1. Tangent spaces

Problem 3.1.3

a.

$$\sigma(t) = \begin{bmatrix} \cos^2(t) - \sin^2(t) & -2\sin(t)\cos(t) & 0 \\ 2\sin(t)\cos(t) & \cos^2(t) - \sin^2(t) & 0 \\ 0 & 0 & 1 \end{bmatrix} = \begin{bmatrix} \cos(2t) & -\sin(2t) & 0 \\ \sin(2t) & \cos(2t) & 0 \\ 0 & 0 & 1 \end{bmatrix}.$$

b. $\sigma(t) \in O(3, \mathbb{R})$ because $O(3, \mathbb{R})$ is closed under multiplication. Directly: $\sigma(t)^T \sigma(t) = I_3$. Also, $\sigma(0) = I_3$.

c. $\sigma'(0) = 2\gamma'(0)$.

Problem 3.1.8

a. We have

$$\gamma(t) = \begin{bmatrix} e^t & 0 \\ 0 & e^{-t} \end{bmatrix} \begin{bmatrix} 1 & t \\ 0 & 1 \end{bmatrix} = \begin{bmatrix} e^t & te^t \\ 0 & e^{-t} \end{bmatrix}.$$

So we see $\det(\beta(t)) = \det(\gamma(t)) = 1$ and $\beta(0) = \gamma(0) = I_2$, so β and γ are smooth curves in $G = SL(2, \mathbb{R})$ passing through the identity. We now have

$$\beta'(t) = \begin{bmatrix} 0 & 1 \\ 0 & 0 \end{bmatrix}, \quad \gamma'(t) = \begin{bmatrix} e^t & te^t + e^t \\ 0 & -e^{-t} \end{bmatrix},$$

so

$$\beta'(0) = \begin{bmatrix} 0 & 1 \\ 0 & 0 \end{bmatrix}, \quad \gamma'(0) = \begin{bmatrix} 1 & 1 \\ 0 & -1 \end{bmatrix}$$

and $\gamma'(0) = \alpha'(0) + \beta'(0)$.

b. Because G is closed under multiplication and $\alpha(t), \beta(t) \in G$, we have $\gamma(t) \in G$. Also, $\gamma(0) = \alpha(0)\beta(0) = I_n I_n = I_n$, so γ gives a (smooth) curve in G passing through the identity. By the product rule, $\gamma'(t) = \alpha(t)\beta'(t) + \alpha'(t)\beta(t)$, so $\gamma'(0) = I_n B + A I_n = A + B$.

c. Because $A + B = \gamma'(0)$ for a smooth curve γ passing through the identity in G, it follows that $A + B \in L(G)$.

Problem 3.1.9

Choose $\boldsymbol{w} \in W$. We know $\boldsymbol{w} = \sum_j a_j \boldsymbol{w}_j$ for some $a_j \in \mathbb{R}$ because the $\boldsymbol{w}_j$ span W. Since $\boldsymbol{w}_j \in U$ for $j = 1, \ldots, m$ and U is closed under addition and scalar multiplication (because U is a subspace), we have $\boldsymbol{w} \in U$. It follows that $W \subseteq U$.

Problem 3.1.11

a. $a(0) = d(0) = 1$ and $b(0) = c(0) = 0$.

b. $(a(t)d(t) - b(t)c(t))' = 1' = 0$ implies $a'(0)d(0) + a(0)d'(0) - b'(0)c(0) - b(0)c'(0) = a'(0) + d'(0) = 0$. So a typical vector $\gamma'(0) \in L(G)$ has trace 0.

c. (i)

$$\begin{bmatrix} a_1 & b_1 \\ c_1 & -a_1 \end{bmatrix} + \begin{bmatrix} a_2 & b_2 \\ c_2 & -a_2 \end{bmatrix} = \begin{bmatrix} a_1 + a_2 & b_1 + b_2 \\ c_1 + c_2 & -(a_1 + a_2) \end{bmatrix} \in W,$$

so W is closed under addition.

$$r \begin{bmatrix} a & b \\ c & -a \end{bmatrix} = \begin{bmatrix} ra & rb \\ rc & -(ra) \end{bmatrix} \in W,$$

so W is closed under scalar multiplication.

(ii) A basis for W is

$$\begin{bmatrix} 1 & 0 \\ 0 & -1 \end{bmatrix}, \begin{bmatrix} 0 & 1 \\ 0 & 0 \end{bmatrix}, \begin{bmatrix} 0 & 0 \\ 1 & 0 \end{bmatrix}.$$

d. Choose

$$\gamma_1(t) = \begin{bmatrix} e^t & 0 \\ 0 & e^{-t} \end{bmatrix}, \gamma_2(t) = \begin{bmatrix} 1 & t \\ 0 & 1 \end{bmatrix}, \gamma_3(t) = \begin{bmatrix} 1 & 0 \\ t & 1 \end{bmatrix}.$$

For $j = 1, 2, 3$, $\det \gamma_j(t) = 1$ and $\gamma_j(0) = I_2$, so each is a (smooth) curve in G passing through I_2. Then $\gamma_1'(0), \gamma_2'(0), \gamma_3'(0)$ are the three basis vectors in (c), and they are in $L(G)$.

e. We know $L(G) \subseteq W$ from (b), and we have shown in (d) that $L(G)$ contains the basis vectors of W, so (3.1.9) implies $W \subseteq L(G)$, and $L(G) = W$ follows.

3.2. Group homomorphisms

Problem 3.2.7.

a. The sum of two rational numbers is a rational number: $\frac{a}{b} + \frac{c}{d} = \frac{ad+bc}{bd}$, so $G_1 = \mathbb{Q}$ is closed under addition. Also 0 is rational ($0 = \frac{0}{1}$ say) and if $a = b/c$ is a nonzero rational number, then $b \neq 0$ and$-a = -b/c$ is also a nonzero rational number.

b. The matrix multiplication below in (c) shows that G_2 is closed under matrix multiplication. The identity matrix is in G_2 (choose $a = 0$ for the upper right corner). Finally, check that

$$\begin{bmatrix} 1 & a \\ 0 & 1 \end{bmatrix}\begin{bmatrix} 1 & -a \\ 0 & 1 \end{bmatrix} = \begin{bmatrix} 1 & 0 \\ 0 & 1 \end{bmatrix} = I_2,$$

so the inverse of any matrix in G_2 exists and is in G_2.

c.

$$F(a)F(b) = \begin{bmatrix} 1 & a \\ 0 & 1 \end{bmatrix}\begin{bmatrix} 1 & b \\ 0 & 1 \end{bmatrix} = \begin{bmatrix} 1 & a+b \\ 0 & 1 \end{bmatrix} = F(a+b).$$

So F is a homomorphism. It's straightforward to check that $F(a) = F(b)$ implies $a = b$, so F is one-to-one. And any matrix $A = (a_{ij}) \in G_2$ has the form $F(a)$ for $a = a_{12} \in G_1$, so F is onto G_2.

d.

$$F^{-1}(A) = F^{-1}\left(\begin{bmatrix} 1 & a \\ 0 & 1 \end{bmatrix}\right) = a,$$

and it's straightforward to check that $F^{-1}(A) + F^{-1}(B) = F^{-1}(AB)$, so F^{-1} is a homomorphism. We see that $F^{-1}(A) = F^{-1}(B)$ implies $a = b$ which in turn implies $A = B$, so F is one-to-one. For every $a \in G_1$, there is a matrix $A \in G_2$ with $F^{-1}(A) = a$ (choose $A = F(a)$), so F is onto G_2.

Problem 3.2.9.

e. $F(I_2) = (I_2^{-1})^T = I_2$.

f. $F(0) = I_2$.

3.3. Differentials

Problem 3.3.1.

a. Write $F(x, y, z) = (x', y', z')$, where $x' = y$, $y' = x$ and $z' = z^2$. For $(x, y, z) \in \mathcal{S}_1$, $z = \sqrt{x^2 + y^2}$, so $z^2 = x^2 + y^2$. This tells us that the points $(x', y', z') \in \mathcal{S}_2$ satisfy $z' = (x')^2 + (y')^2$ and $\mathcal{S}_2$ is a paraboloid (an upward opening bowl).

b. First note that $\gamma_1(t) \in \mathcal{S}_1$, since $t = \sqrt{t^2 + 0^2}$ for $t \geq 0$. The curve parameterized by γ_1 is the ray on the cone $\mathcal{S}_1$ sitting above the positive x axis — in other words it's the ray $z = x, x \geq 0$ in the xz-plane. Choose $t_0 = 3$. We have $\gamma_1'(t) = (1, 0, 1) = \gamma_1'(t_0)$. The vector $\gamma_1'(3)$ lies along the line $z = x$ in the xz-plane and points up with slope 1—i.e., since the curve given by γ_1 is along a straight line, the tangent vector lies on the "curve".

c. $\gamma_2(t) = F(t, 0, t) = (0, t, t^2)$. This curve follows the parabola $z = y^2$ in the yz-plane; to describe it another way, it is the curve following the paraboloid and lying above the positive y axis. $\gamma_2(t_0) = \gamma_2(3) = (0, 3, 9)$. $\gamma_2'(t) = (0, 1, 2t)$, so $\gamma_2'(3) = (0, 1, 6)$. This vector is tangent to the parabola at (0, 3, 9) in the yz-plane and points up with slope 6.

Problem 3.3.2.

a. We have $f_1(x, y, z) = y$, $f_2(x, y, z) = x$ and $f_3(x, y, z) = z^2$, so the Jacobian matrix of F is

$$\begin{bmatrix} 0 & 1 & 0 \\ 1 & 0 & 0 \\ 0 & 0 & 2z \end{bmatrix}.$$

Evaluating at $\boldsymbol{p}_1 = (3, 0, 3)$ gives

$$\begin{bmatrix} 0 & 1 & 0 \\ 1 & 0 & 0 \\ 0 & 0 & 6 \end{bmatrix}.$$

Finally,

$$\begin{bmatrix} 0 & 1 & 0 \\ 1 & 0 & 0 \\ 0 & 0 & 6 \end{bmatrix} \begin{bmatrix} 1 \\ 0 \\ 1 \end{bmatrix} = \begin{bmatrix} 0 \\ 1 \\ 6 \end{bmatrix} = \gamma_2'(3).$$

b. In general, $\gamma_2(t) = F(a(t), b(t), c(t))$. Calculating with the Jacobian matrix at $\gamma_1(t)$ we get

$$\begin{bmatrix} 0 & 1 & 0 \\ 1 & 0 & 0 \\ 0 & 0 & 2c(t) \end{bmatrix} \begin{bmatrix} a'(t) \\ b'(t) \\ c'(t) \end{bmatrix} = \begin{bmatrix} b'(t) \\ a'(t) \\ 2c(t)c'(t) \end{bmatrix}.$$

The multivariable chain rule says the jth component of $\gamma_2'(t) = \frac{d}{dt} F(\gamma_1(t))$ is

$$\frac{d}{dt} f_j(a(t), b(t), c(t)) = \frac{\partial f_j}{\partial x}\frac{dx}{dt} + \frac{\partial f_j}{\partial y}\frac{dy}{dt} + \frac{\partial f_j}{\partial z}\frac{dz}{dt}.$$

This tells us

$$\frac{d}{dt}f_1(a(t),b(t),c(t)) = 1\cdot b'(t),$$
$$\frac{d}{dt}f_2(a(t),b(t),c(t)) = 1\cdot a'(t),$$
$$\frac{d}{dt}f_3(a(t),b(t),c(t)) = 2c(t)c'(t),$$

which agrees with the calculation using the Jacobian matrix.

Problem 3.3.4.

The Jacobian matrix of F is

$$\begin{bmatrix} \sin(x_2) & x_1\cos(x_2) \\ x_2\cos(x_1x_2) & x_1\cos(x_1x_2) \\ 2x_1 & 1 \end{bmatrix} \text{ which equals } \begin{bmatrix} 0 & 0 \\ \pi & 0 \\ 0 & 1 \end{bmatrix}$$

at $x_1 = 0$, $x_2 = \pi$.

b.

$$\begin{bmatrix} 0 & 0 \\ \pi & 0 \\ 0 & 1 \end{bmatrix}\begin{bmatrix} 1 \\ 3 \end{bmatrix} = \begin{bmatrix} 0 \\ \pi \\ 3 \end{bmatrix}.$$

Problem 3.3.5.

a. The matrix is

$$\begin{bmatrix} 0 & 1 & 0 \\ 1 & 0 & 3 \end{bmatrix}.$$

b. The result is $(0.1, 0.2)$ (writing it as a row vector).

c. To four decimal places, $(0.9)^2\sin(0,1) = 0.0809$, and $(0.9)(1.1)^3 = 1.1979$, so $F(0.9, 0.1, 1.1) - F(1,0,1) = (0.0809, 0.1979) \approx (0.1, 0.2)$ to one decimal place.

Problem 3.3.6.

b. The matrix of dF at $(1,0,0,1)$ is

$$\begin{bmatrix} 0 & 0 & 0 & 1 \\ 0 & 0 & -1 & 0 \\ 0 & -1 & 0 & 0 \\ 1 & 0 & 0 & 0 \end{bmatrix}.$$

Problem 3.3.8.

a. The matrix of dF has one row and four columns. It is $[x_{22}, -x_{21}, -x_{12}, x_{11}]$. Evaluating at $I_2 = (1,0,0,1)$ we get $dF = [1,0,0,1]$. Multiplying this matrix dF by the column vector representing the matrix X we get $dF(X) = x_{11} + x_{22} = \text{tr}(X)$.

4. One-parameter subgroups and the exponential map

4.1. One-parameter subgroups

Problem 4.1.1.

In each case, $\det(\gamma(t)) = 1 \neq 0$, so $\gamma: \mathbb{R} \to GL(2, \mathbb{R})$.

c.

$$\gamma(t)\gamma(s) = \begin{bmatrix} e^t & 0 \\ 0 & e^{-t} \end{bmatrix}\begin{bmatrix} e^s & 0 \\ 0 & e^{-s} \end{bmatrix} = \begin{bmatrix} e^{t+s} & 0 \\ 0 & e^{-(t+s)} \end{bmatrix} = \gamma(t+s).$$

4.2. The exponential map in dimension 1

Problem 4.2.1.

Assume $\gamma(1) = b < 0$. Then we have $b < 0 < 1$ with $\gamma(1) = b$ and $\gamma(0) = 1$. Since γ is a continuous function, the Intermediate Value Theorem implies there is a t between 0 and 1 for which $\gamma(t) = 0$. But this contradicts $\gamma(t) \neq 0$ for all t. So the assumption $b < 0$ must be false. We can't have $b = 0$ since we know $\gamma(1) \neq 0$, so it must be that $b > 0$.

Problem 4.2.2.

a. The statement is trivially true for $m = 1$. Suppose the statement is true for some $m = k \geq 1$: $\gamma(ka) = (\gamma(a))^k$. Then $\gamma((k+1)a) = \gamma(ka + a) = \gamma(ka)\gamma(a)$, since γ is a homomorphism. But this equals $\gamma(a)^k\gamma(a)$ by the induction hypothesis. And this last expression equals $\gamma(a)^{k+1}$, which means the truth of the statement for $m = k$ implies the truth of the statement for $m = k + 1$. Therefore, by induction, the statement is true for all positive integers m.

c. Since γ is a homomorphism, $1 = \gamma(0) = \gamma(m + (-m)) = \gamma(m)\gamma(-m)$. Solve the equation $1 = b^m\gamma(-m)$ for $\gamma(-m) = 1/b^m = b^{-m}$.

4.3. Calculating the matrix exponential

Problem 4.3.2.

b.

$$\exp(tA) = \begin{bmatrix} e^t & 0 \\ 0 & e^{-t} \end{bmatrix}.$$

Problem 4.3.3.

b.

$$\exp(tA) = \begin{bmatrix} \cos(t) & -\sin(t) \\ \sin(t) & \cos(t) \end{bmatrix}.$$

4.4. Properties of the matrix exponential

Problem 4.4.1.

a. $\exp(0_n) = I_n + 0_n + 0_n + \cdots = I_n$.

b. Since A and $-A$ commute, $I_n = \exp(0_n) = \exp(A + (-A)) = \exp(A)\exp(-A)$, so $\exp(-A)$ is the inverse of $\exp(A)$, which is therefore invertible.

c. The statement is true for $m = 1$. Assume $(A^k)^T = (A^T)^k$ for some $m = k \geq 1$. Then $(A^{k+1})^T = (A^k A)^T = A^T (A^k)^T = A^T (A^T)^k = (A^T)^{k+1}$, so the statement holds for $m = k + 1$. Therefore, by induction, it holds for all positive integers m.

d. Using the fact that the transpose of a sum is the sum of the transposes, (c) shows that the series for $\exp(A^T)$ is the transpose of the series for $\exp(A)$.

Problem 4.4.3.

c. $[(t^m/m!)A^m]' = (t^m/m!)' A^m = (t^{m-1}/(m-1)!)A^m$, so

$$\gamma'(t) = (\exp(tA))' = A + tA^2 + \cdots + \frac{t^{m-1}}{(m-1)!}A^m + \cdots = A\exp(tA).$$

Since $\exp(0_n) = I_n$, it follows that $\gamma'(0) = AI_n = A$.

4.5. Using exp to determine $L(G)$

Problem 4.5.3.

a. Using the product rule for matrices and the fact that $(\gamma'(t))^T = (\gamma(t)^T)'$ from (2.1.12), we have $(\gamma'(t))^T C\gamma(t) + \gamma(t)^T C\gamma'(t) = 0_2$ and therefore $\gamma'(0)^T C + C\gamma'(0) = 0_2$. Assume the ij-entry of $\gamma'(0)$ is a_{ij}. Then we have

$$\begin{bmatrix} a_{11} & a_{21} \\ a_{12} & a_{22} \end{bmatrix}\begin{bmatrix} 1 & 0 \\ 0 & -1 \end{bmatrix} + \begin{bmatrix} 1 & 0 \\ 0 & -1 \end{bmatrix}\begin{bmatrix} a_{11} & a_{12} \\ a_{21} & a_{22} \end{bmatrix} = \begin{bmatrix} 0 & 0 \\ 0 & 0 \end{bmatrix}$$

which implies $a_{11} = a_{22} = 0$ and $a_{12} = a_{21}$. But this says $\gamma'(0) \in W$. Since every element of $L(G)$ has the form $\gamma'(0)$ for such a γ, this tells us $L(G) \subseteq W$.

b. Calculating from the definition of $\exp(tB)$ we find

$$\exp(tB) = \begin{bmatrix} \cosh(tb) & \sinh(tb) \\ \sinh(tb) & \cosh(tb) \end{bmatrix}.$$

(Compare (4.3.4).) We recognize this as the matrix of a Lorentz rotation of the first kind.

Problem 4.5.4.

b. Let $B_1, B_2 \in W$, so $\mathrm{tr}(B_j) = 0$ for $j = 1, 2$. Then $\mathrm{tr}(B_1 + B_2) = \mathrm{tr}(B_1) + \mathrm{tr}(B_2) = 0 + 0 = 0$, so $B_1 + B_2 \in W$. Also $\mathrm{tr}(rB) = r\ \mathrm{tr}(B) = 0$, so $rB \in W$ for all $r \in \mathbb{R}$.

4.6. Differential equations

Problem 4.6.1

If we write $\gamma(t) = \exp(tA)$, then $d\mathbf{y}/dt = \gamma'(t)\mathbf{x}$ by the product rule (since $d\mathbf{x}/dt = \mathbf{0}$), so $d\mathbf{y}/dt = A\gamma(t)\mathbf{x} = A\mathbf{y}$. Also, evaluating at $t = 0$, we get $\mathbf{y}(0) = \gamma(0)\mathbf{x} = I_n\mathbf{x} = \mathbf{x}$.

Problem 4.6.2

a. Using (4.3.2),

$$A = \begin{bmatrix} 1 & 0 \\ 0 & -1 \end{bmatrix} \text{ and } \exp(tA) = \begin{bmatrix} e^t & 0 \\ 0 & e^{-t} \end{bmatrix},$$

which gives $y_1 = x_1 e^t$ and $y_2 = x_2 e^{-t}$.

b. $dy_1/dt = x_1 e^t = y_1$, and $dy_2/dt = -x_2 e^{-t} = -y_2$. Also, $y_1(0) = x_1$ and $y_2(0) = x_2$.

Problem 4.6.9

a.

$$A = \begin{bmatrix} 1 & 3 \\ 3 & 1 \end{bmatrix} \text{ and } \exp(tA) = (1/2)\begin{bmatrix} e^{4t} + e^{-2t} & e^{4t} - e^{-2t} \\ e^{4t} - e^{-2t} & e^{4t} + e^{2t} \end{bmatrix},$$

which gives $y_1 = (x_1/2)(e^{4t} + e^{-2t}) + (x_2/2)(e^{4t} - e^{-2t})$ and $y_2 = (x_1/2)(e^{4t} - e^{-2t}) + (x_2/2)(e^{4t} + e^{-2t})$.

5. Lie algebras

5.1. Lie algebras

Problem 5.1.2.

a. $AB - BA \in \mathcal{M}(n, \mathbb{R})$ if $A, B \in \mathcal{M}(n, \mathbb{R})$.

b. $[B, A] = BA - AB = -(AB - BA) = -[A, B]$ for all $A, B \in \mathcal{M}(n, \mathbb{R})$.

c. For all $A, B, C \in \mathcal{M}(n, \mathbb{R})$ and all $r \in \mathbb{R}$, $[A+B, C] = (A+B)C - C(A+B) = AC + BC - CA - CB = AC - CA + BC - CB = [A, C] + [B, C]$, and similarly for $[A, B+C]$. Also, $r[A, B] = r(AB - BA) = r(AB) - r(BA) = (rA)B - B(rA) = [rA, B] = [A, rB]$.

d. $[A, [B, C]] + [B, [C, A]] + [C, [A, B]] = A(BC - CB) - (BC - CB)A + B(CA - AC) - (CA - AC)B + C(AB - BA) - (AB - BA)C = ABC - ACB + BCA + CBA + BCA - BAC - CAB + ACB + CAB - CBA - ABC + BAC = 0_n$ for all $A, B, C \in \mathcal{M}(n, \mathbb{R})$.

Problem 5.1.5.

$$\left[\begin{bmatrix} 0 & b_1 \\ b_1 & 0 \end{bmatrix}, \begin{bmatrix} 0 & b_2 \\ b_2 & 0 \end{bmatrix}\right] = 0_2 \in L(G).$$

(In fact, $L(G)$ is abelian in this case.)

Problem 5.1.7.

Assume $A, B \in L(G)$, so $\mathrm{tr}(A) = \mathrm{tr}(B) = 0$. Then $\mathrm{tr}[A, B] = \mathrm{tr}(AB - BA) = \mathrm{tr}(AB) - \mathrm{tr}(BA) = 0$, so $[A, B] \in L(G)$. (This is easier than the direct calculation in (5.1.5).)

Problem 5.1.10.

b.

$$\text{Let } W = \left\{ \begin{bmatrix} 0 & u & v \\ 0 & 0 & w \\ 0 & 0 & 0 \end{bmatrix} : u, v, w \in \mathbb{R} \right\}.$$

Any curve $\gamma(t)$ in $\mathcal{H}$ must have constant entries in all except the 12, 13, and 23-entries, so $\gamma'(0)$ must have zeroes in all but possibly some of these entries. Therefore, $L(\mathcal{H}) \subseteq W$. To show the reverse inclusion, consider the curves

$$\gamma_1(t) = \begin{bmatrix} 1 & t & 0 \\ 0 & 1 & 0 \\ 0 & 0 & 1 \end{bmatrix}, \gamma_2(t) = \begin{bmatrix} 1 & 0 & t \\ 0 & 1 & 0 \\ 0 & 0 & 1 \end{bmatrix}, \gamma_3(t) = \begin{bmatrix} 1 & 0 & 0 \\ 0 & 1 & t \\ 0 & 0 & 1 \end{bmatrix}.$$

Clearly these are smooth curves in $\mathcal{H}$ passing through the identity. The vectors $A_j = \gamma_j'(0)$ for $j = 1, 2, 3$ form a basis for W, and each is in $L(\mathcal{H})$, so by (3.1.9), $L(\mathcal{H}) = W$.

Problem 5.1.12.

c.

$$T(\boldsymbol{v}_1 + \boldsymbol{v}_2) = \begin{bmatrix} 0 & -x_1 - x_2 & -y_1 - y_2 \\ x_1 + y_1 & 0 & -z_1 - z_2 \\ y_1 + y_2 & z_1 + z_2 & 0 \end{bmatrix} = T(\boldsymbol{v}_1) + T(\boldsymbol{v}_2).$$

Also,

$$T(r\boldsymbol{v}) = \begin{bmatrix} 0 & -rx & -ry \\ rx & 0 & -rz \\ ry & rz & 0 \end{bmatrix} = rT(\boldsymbol{v}).$$

So T is a linear transformation. If $T(\boldsymbol{v}_1) = T(\boldsymbol{v}_2)$ then, by comparing matrix entries, we get $x_1 = x_2$, $y_1 = y_2$, $z_1 = z_2$, so $\boldsymbol{v}_1 = \boldsymbol{v}_2$ and T is one-to-one. And every anti-symmetric 3 by 3 matrix has a pre-image in $\mathbb{R}^3$, so T is onto. (Equivalently, define T^{-1} in the obvious way.) Finally, we need to show T preserves the bracket: $T([\boldsymbol{v}, \boldsymbol{w}]) = [T(\boldsymbol{v}), T(\boldsymbol{w})]$. First, $\boldsymbol{v} \times \boldsymbol{w} = (v_2w_3 - w_2v_3, -v_1w_3 + w_1v_3, v_1w_2 - w_1v_2)$, so

$$T(\boldsymbol{v} \times \boldsymbol{w}) = \begin{bmatrix} 0 & -(v_2w_3 - w_2v_3) & -(-v_1w_3 + w_1v_3) \\ v_2w_3 - w_2v_3 & 0 & -(v_1w_2 - w_1v_2) \\ -v_1w_3 + w_1v_3 & v_1w_2 - w_1v_2 & 0 \end{bmatrix}.$$

On the other hand, $[T(\boldsymbol{v}), T(\boldsymbol{w})] = T(\boldsymbol{v})T(\boldsymbol{w}) - T(\boldsymbol{w})T(\boldsymbol{v}) =$

$$\begin{bmatrix} 0 & -v_1 & -v_2 \\ v_1 & 0 & -v_3 \\ v_2 & v_3 & 0 \end{bmatrix} \begin{bmatrix} 0 & -w_1 & -w_2 \\ w_1 & 0 & -w_3 \\ w_2 & w_3 & 0 \end{bmatrix} - \begin{bmatrix} 0 & -w_1 & -w_2 \\ w_1 & 0 & -w_3 \\ w_2 & w_3 & 0 \end{bmatrix} \begin{bmatrix} 0 & -v_1 & -v_2 \\ v_1 & 0 & -v_3 \\ v_2 & v_3 & 0 \end{bmatrix},$$

and carrying out the multiplications and simplifying gives the same result as $T(\boldsymbol{v} \times \boldsymbol{w}) = T([\boldsymbol{v}, \boldsymbol{w}])$.

5.2. Adjoint maps—big 'A' and small 'a'

Problem 5.2.1

Let $X, Y \in \mathcal{M}(n, \mathbb{R})$ and $r \in \mathbb{R}$.

a. $\mathrm{Ad}\,(M)(X+Y) = M(X+Y)M^{-1} = (MX+MY)M^{-1} = MXM^{-1}+MYM^{-1} = \mathrm{Ad}\,(M)(X) + \mathrm{Ad}\,(M)(Y)$, and $\mathrm{Ad}\,(M)(rX) = M(rX)M^{-1} = rMXM^{-1} = r\,\mathrm{Ad}\,(M)(X)$.

b. $\mathrm{Ad}\,(M)[X,Y] = M(XY - YX)M^{-1} = M(XY)M^{-1} - M(YX)M^{-1} = (MXM^{-1})(MYM^{-1}) - (MYM^{-1})(MXM^{-1}) = [\mathrm{Ad}\,(M)(X), \mathrm{Ad}\,(M)(Y)]$, so $\mathrm{Ad}\,(M)$ preserves the matrix bracket on $\mathcal{M}(n, \mathbb{R})$.

c. The inverse of $\mathrm{Ad}\,(M)$ is $\mathrm{Ad}\,(M^{-1})$ since $M^{-1}(MXM^{-1})(M^{-1})^{-1} = X$ for all $X \in \mathcal{M}(n, \mathbb{R})$, so $\mathrm{Ad}\,(M^{-1}) \circ \mathrm{Ad}\,(M) = I$, the identity function (and $\mathrm{Ad}\,(M) \circ \mathrm{Ad}\,(M^{-1}) = I$ also).

Problem 5.2.2

For $X \in \mathcal{M}(n, \mathbb{R})$, $\mathrm{Ad}(M_1 M_2)(X) = (M_1 M_2)X(M_1 M_2)^{-1} = (M_1 M_2)X(M_2^{-1} M_1^{-1})$ by (3.2.5b). This last expression is the same as

$$M_1(M_2 X M_2^{-1})M_1^{-1} = \mathrm{Ad}\,(M_1)(\mathrm{Ad}\,(M_2)(X)) = (\mathrm{Ad}\,(M_1) \circ \mathrm{Ad}\,(M_2))(X).$$

So $\mathrm{Ad}(M_1 M_2) = \mathrm{Ad}(M_1)\circ\, \mathrm{Ad}(M_2)$.

Problem 5.2.7

a. $[E, H] = -2E$, $[F, H] = 2F$, $[E, F] = H$.

c.

$$\mathcal{E} = \begin{bmatrix} 0 & -2 & 0 \\ 0 & 0 & 1 \\ 0 & 0 & 0 \end{bmatrix}.$$

e.

$$\exp(\mathcal{E}) = \begin{bmatrix} 1 & -2 & -1 \\ 0 & 1 & 1 \\ 0 & 0 & 1 \end{bmatrix}, \exp(\mathcal{H}) = \begin{bmatrix} e^2 & 0 & 0 \\ 0 & 1 & 0 \\ 0 & 0 & e^{-2} \end{bmatrix}, \exp(\mathcal{F}) = \begin{bmatrix} 1 & 0 & 0 \\ -1 & 1 & 0 \\ -1 & 2 & 1 \end{bmatrix}.$$

Problem 5.2.8

b.

$$M_E = \begin{bmatrix} 1 & 1 \\ 0 & 1 \end{bmatrix}, M_F = \begin{bmatrix} 1 & 0 \\ 1 & 1 \end{bmatrix}, M_H = \begin{bmatrix} e & 0 \\ 0 & e^{-1} \end{bmatrix}.$$

c. $\mathrm{Ad}(M_E) = \exp(\mathcal{E})$

6. Matrix groups over other fields

6.1. What is a field?

Problem 6.1.2

Note that these arguments are written quite abstractly, tying every calculation to the axioms and definitions. Thus these arguments are valid in *any* field, not just the familiar ones. By the distributive law and the definition of 0, $0a = (0 + 0)a = 0a + 0a$. We know $\mathbb{F}$

is closed under multiplication, so $0a \in \mathbb{F}$, and its additive inverse is in $\mathbb{F}$. Adding the additive inverse of $0a$ to both sides of $0a = 0a + 0a$ gives $-0a + 0a = -0a + (0a + 0a)$. Using the associative law for addition and the definitions of additive inverse and additive identity, this implies $0 = 0a$. Using the distributive law and the definition of additive inverse, $0 = 0a = (1 + (-1))a = 1a + (-1)a$. But $1a = a$ by the definition of multiplicative identity, so this gives $0 = a + (-1)a$, which tells us that $(-1)a$ is the additive inverse of a; i.e., $(-1)a = -a$.

6.2. The unitary group

Problem 6.2.2

d. Let $\boldsymbol{e}_j$ be the vector with jth component 1 and others 0. Then $\boldsymbol{v} \cdot \boldsymbol{e}_j = 0$ for all j. But then $\boldsymbol{v} \cdot \boldsymbol{e}_j = \overline{v}_j = 0$, for $j = 1, \ldots, n$, which implies $v_j = 0$ for all j, so $\boldsymbol{v} = \boldsymbol{0}$.

Problem 6.2.6

a. $\gamma(t) \in U(n, \mathbb{C})$ implies $\overline{\gamma(t)}^T \gamma(t) = I_n$. Differentiating both sides and using (6.2.4) gives $\overline{\gamma(t)}^T \gamma'(t) + \overline{\gamma'(t)}^T \gamma(t) = 0_n$. Evaluating at $t = 0$ and using $\gamma(0) = I_n$ gives $\overline{\gamma'(0)}^T + \gamma'(0) = 0_n$.

Problem 6.2.7

a. $\exp(tA) = \begin{bmatrix} e^{it} & 0 \\ 0 & 1 \end{bmatrix}$ **b.** $\exp(tA) = \begin{bmatrix} 1 & it \\ 0 & 1 \end{bmatrix}$.

Problem 6.2.8

a.

$$M_1 = \exp(tA_1) = \begin{bmatrix} \cos(t) & i\sin(t) \\ i\sin(t) & \cos(t) \end{bmatrix}.$$

Since $\overline{M_1}^T M_1 = I_2$ and $\det(M_1) = 1$, $M_1 \in SU(2, \mathbb{C})$.

6.3. Matrix groups over finite fields

Problem 6.3.2

a.

+	0	1	2	3	4
0	0	1	2	3	4
1	1	2	3	4	0
2	2	3	4	0	1
3	3	4	0	1	2
4	4	0	1	2	3

and

·	0	1	2	3	4
0	0	0	0	0	0
1	0	1	2	3	4
2	0	2	4	1	3
3	0	3	1	4	2
4	0	4	3	2	1

b. Both operations are commutative since the tables are symmetric. The addition table shows $\mathbb{F}_5$ is closed under addition and contains the additive identity 0. Additive inverses are $-0 = 0$, $-1 = 4$, $-2 = 3$, $-3 = 2$, $-4 = 1$. So $\mathbb{F}_5$ is an abelian group under addition. The multiplication table shows the set of nonzero elements of $\mathbb{F}_5$ is closed under multiplication and contains the multiplicative identity 1. Multiplicative inverses are $1^{-1} = 1$, $2^{-1} = 3$, $3^{-1} = 2$, $4^{-1} = 4$. So the nonzero elements of $\mathbb{F}_5$ form an abelian group under multiplication.

c. $3 - 4 = 3 + (-4) = 3 + 1 = 4$. $3/4 = 3 \cdot 4^{-1} = 3 \cdot 4 = 2$.

d. The integer 5 is $1 + 1 + 1 + 1 + 1 + 1 = 4 + 1 = 0$. The integer 17 is $1 + \cdots + 1$ (17 terms), and we can rearrange this as a sum of 5 terms plus 5 terms plus 5 terms plus $1 + 1$. But each sum of 5 terms equals 0, so 17 is $0 + 0 + 0 + 1 + 1 = 2$.

Problem 6.3.5

a. Writing vectors as rows and without commas or parentheses, the nonzero vectors of V are $100, 010, 001, 110, 101, 011, 111$.

b. With the convention of (a), the nonzero vectors in the two-dimensional subspaces of V are

$100, 010, 110$
$100, 001, 101$
$100, 011, 111$
$010, 001, 011$
$010, 111, 101$
$001, 110, 111$
$011, 101, 110$

Problem 6.3.7

a. $\det(M) = 1$.

b. $M_1^2 = I_3$.

c. T_1 is a reflection with mirror line the angle bisector at vertex 100. It fixes 100, 111 and 011 on the mirror line. It interchanges the other two vertices 110 and 101 and also interchanges the midpoints 010 and 001.

Bibliography

Artin, E., *Geometric Algebra*, Interscience Tracts in Pure and Applied Mathematics **3**, John Wiley and Sons (1957).

Artin, M., *Algebra*, Prentice Hall (1991).

Baez, J., "Octonions," *Bulletin of the American Mathematical Society* **39** (2002), pp. 145–205. (http://math.ucr.edu/home/baez/octonions.)

Brown, E., "The many names of $(7, 3, 1)$," *Mathematics Magazine* **75** (2002) pp. 83–94.

Brown, E.,"The Fabulous $(11, 5, 2)$ Biplane," *Mathematics Magagazine* **77** (2004), 87–100.

Brown, E. and Mellinger, K.E., "Kirkman's Schoolgirls Wearing Hats and Walking through Fields of Numbers," *Mathematics Magazine* **82** (2009) pp. 3–15.

Burn, R.P. *Groups: A Path to Geometry*, Cambridge University Press (1985).

Callahan, J.J., *The Geometry of Spacetime: An Introduction to Special and General Relativity*, Springer-Verlag (2000).

Carter, R.W., *Simple Groups of Lie Type*, Interscience Texts and Monographs in Pure and Applied Mathematics **28**, John Wiley and Sons (1972).

J.H. Conway and D.A. Smith, *On Quaternions and Octonians : Their Geometry, Arithmetic, and Symmetry*, A.K. Peters (2003).

Cromwell, P.R., *Polyhedra*, Cambridge University Press (1997).

Curtis, M.L., *Matrix Groups*, Universitext, Springer-Verlag (1979).

Dray, T. and C. Manogue, "The Octonionic Eigenvalue Problem," *Adv. Appl. Clifford Algebras* **8** (1998), pp. 241–364. (arXiv.math/9807126 v1 (1998))

Farmer, D. W., *Groups and Symmetry: A Guide to Discovering Mathematics*, Mathematical World **5**, American Mathematical Society (1996).

Gorenstein, D., *Finite Groups*, Harper and Row (1968).

Hausner, M. and Schwartz, J.T., *Lie Groups; Lie Algebras*, Notes on Mathematics and its Applications, Gordon and Breach (1968).

Howe, R., "Very Basic Lie Theory," *American Mathematical Monthly* **90**, no. 9 (1983), pp. 600–623.

Howe, R. and Barker, W., *Continuous Symmetry: From Euclid to Einstein*, manuscript (2000). (The first volume, consisting of chapters I-IX, has been published by the American Mathematical Society, 2007.)

Humphreys, J.E., *Introduction to Lie Algebras and Representation Theory*, Graduate Texts in Mathematics, Springer-Verlag (1972).

Ise, M. and Takeuchi, M., *Lie Groups I*, Translations of Mathematical Monographs **85**, American Mathematical Society (1991).

Lander, E., *Symmetric Designs: an Algebraic Approach*, London Mathematical Society Lecture Note Series **74**, Cambridge University Press (1983).

Mumford, D. *et al*, *Indra's Pearls: The Vision of Felix Klein*, Cambridge University Press (2002).

Nielsen, M. and Chuang, I., *Quantum Computation and Quantum Information*, Cambridge University Press (2000).

Pollatsek, H., "Quantum error correction: classic group theory meets a quantum challenge," *American Mathematical Monthly* **108**, no. 10 (2001), pp. 932–962.

Reid, C., *Hilbert*, Copernicus, Springer-Verlag (1996).

Ronan, M., *Symmetry and the Monster*, Oxford University Press (2006).

Rossmann, W. *Lie Groups: An Introduction through Linear Groups*, Oxford University Press (2002).

Sattinger, D.H. and Weaver, O.L., *Lie Groups and Algebras with Applications to Physics, Geometry and Mechanics*, Applied Mathematical Sciences **61**, Springer-Verlag (1986).

Singer, S.F., *Linearity, Symmetry, and Prediction in the Hydrogen Atom*, Undergraduate Texts in Mathematics, Springer-Verlag (2005).

Singer, S.F., *Symmetry in Mechanics: A Gentle, Modern Introduction*, Birkhäuser (2001, second edition 2004).

Snapper, E. and Troyer, R.J., *Metric Affine Geometry*, Holden-Day (1968).

Stillwell, J., *Naive Lie Theory*, Undergraduate Texts in Mathematics, Springer-Verlag (2008).

Tapp, K., "Matrix Groups for Undergraduates," Student Mathematical Library **29**, American Mathematical Society (2005).

Taylor, E.F. and Wheeler, J.A., *Spacetime Physics*, W.H. Freeman (1966, second edition 1992).

Thompson, T.M., *From Error-Correcting Codes through Sphere Packings to Simple Groups*, Carus Mathematical Monographs **21**, Mathematical Association of America (1983).

Yaglom, I.M. (translated by S. Sossensky, edited by H. Grant and A. Shenitzer), *Felix Klein and Sophus Lie: Evolution of the Idea of Symmetry in the Nineteenth Century*, Birkhäuser (1988).

Yale, P.B. *Geometry and Symmetry*, Holden-Day (1968). (Republished as a Dover Phoenix edition, 2004.)

Zulli, L., "Charting the 3-sphere—an exposition for undergraduates," *American Mathematical Monthly* **103**, no. 3 (1996), 221–229.

Index

Notation

$\|A\|$ length of matrix A 93
ad adjoint map (small a) 105, 107, 109
Ad Adjoint map (big A) 104, 109
Aut(L) automorphism group of L 132

$\mathcal{B}_r(A)$ open ball around matrix A 96

$\mathbb{C}$ complex numbers 29

dim(G) dimension of G 66

$e^{i\alpha}$ 31
exp(A) matrix exponential 72, 87

$\mathbb{F}_2$, $\mathbb{F}_q$ (q power of a prime) finite field 121, 122, 127, 128

$GF(q)$ Galois field 128
$G(L, \mathbb{F})$ Chevalley group of adjoint type 135
$GL(n, \mathbb{R})$ general linear group (matrices) 9, 23
$GL(\mathbb{R}^n)$ general linear group (transformations) 9, 22

$\mathbb{H}$ quaternions 33

$\mathcal{L}(2, \mathbb{R})$, $\mathcal{L}(4, \mathbb{R})$ Lorentz group (matrices) 23, 24
$\mathcal{L}(\mathbb{R}^2)$, $\mathcal{L}(\mathbb{R}^4)$ Lorentz group (transformations) 16, 22
$\mathcal{L}^{++}(2, \mathbb{R})$ Lorentz rotations of first kind (matrices) 18
$L(G)$ tangent space, Lie algebra 46, 109

log(M) matrix logarithm 91
$\mathcal{M}(n, \mathbb{C})$ (matrices) 32
$\mathcal{M}(n, \mathbb{R})$ (matrices) 9

$\mathbb{O}$ octonions 41
$O(2, \mathbb{R})$, $O(n, \mathbb{R})$ orthogonal group (matrices) 10, 12, 24
$O(\mathbb{R}^2)$, $O(\mathbb{R}^n)$ orthogonal group (transformations) 10, 22

$PSL(2, \mathbb{F})$ projective special linear group 136

S^1, S^2, S^3 spheres 31, 34, 39
$SL(n, \mathbb{R})$ special linear group (matrices) 10, 23
$SL(\mathbb{R}^n)$ special linear group (transformations) 10
$SO(2, \mathbb{R})$, $SO(n, \mathbb{R})$ rotation group (matrices) 10, 12, 24
$SO(\mathbb{R}^n)$ rotation group (transformations) 10, 22
$Sp(n, \mathbb{R})$ (n even) symplectic group (matrices) 24
$Sp(\mathbb{R}^n)$ (n even) symplectic group (transformations) 22
$SU(n, \mathbb{C})$ special unitary group (matrices) 32, 39, 116

$U(n, \mathbb{C})$ unitary group (matrices) 32, 39, 116
$U(\mathbb{C}^n)$ unitary group (transformations) 118, 129

(v, k, λ) difference set 126
(v, k, λ) symmetric design 124

$\overline{z}$ conjugate of z
  complex 29, 39
  quaternionic 34, 40
$|z|$ length of z
  for z complex 30
  for z quaternion 34

About the Author

Harriet Pollatsek was a mathematics major at the University of Michigan and stayed on for her PhD. Her thesis (with J.E. McLaughlin) and early papers were on finite classical groups and their geometries. She has also done research on difference sets (directing two NSF-funded undergraduate research groups) and quantum error-correction. She is the Julia and Sarah Ann Adams Professor of Mathematics at Mount Holyoke College, where she has been since 1970, except for sabbatical leaves at the Universities of Oregon, Cambridge, Sussex and London (Queen Mary College).

Curriculum development and writing with colleagues are longtime interests. Projects include the interdisciplinary *Case Studies in Quantitative Reasoning* (funded by the A.P. Sloan Foundation), the Five College *Calculus in Context* (National Science Foundation), *Laboratories in Mathematical Experimentation* (NSF), *Increasing Access to Advanced Mathematics* (Fund for the Improvement of Post-Secondary Education), and *Mathematics Across the Curriculum* (NSF and the National Endowment for the Humanities). She has been awarded the Mount Holyoke College Teaching Award and the Association for Women in Mathematics' Louise Hay Award for contributions to mathematics education.

She is a member of the Mathematical Association of America, the American Mathematical Society and the Association for Women in Mathematics. She was chair of the MAA's Committee on the Undergraduate Program in Mathematics and led the writing team for *Undergraduate Programs and Courses in the Mathematical Sciences: CUPM Curriculum Guide 2004*. She now chairs the MAA's Council on Programs and Students in the Mathematical Sciences.

She and her husband Alexander (Sandy) have two children: Shura is an assistant professor of costume design at the University of Western Kentucky, and David is a vice president at Monster Games in Northfield, Minnesota. Harriet and Sandy have lived in Amherst, Massachusetts, since 1969.